Table of Contents

About the National Science and Technology Council 2

About This Report 2

Executive Overview. 3

The National Biometrics Challenge 5

1. Introduction 5

2. Challenges Met and Remaining Since 2006. 7
   2.1 Interoperability 7
   2.2 Technology 9
   2.3 Notable Successes 15

3. The 2011 Environment 18
   3.1 Technology Landscape 18
   3.2 Cost and Institutional Constraints 21
   3.3 Privacy, Civil Rights and Civil Liberties Protection 23
   3.4 Interagency Cooperation and Partnering 25

4. User-Centric Biometrics Approach 26

5. Science and Technology 27
   5.1 Research Challenges 27
   5.2 Research Focus Areas 27
   5.3 Education and Training. 31

6. Standards, Commercial Testing and Certification 33
   6.1 Accomplishments 34
   6.2 Trends and Challenges. 36

7. Conclusion. 38

NSTC Subcommittee on Biometrics and Identity Management 39

Department Leads. 39

Research, Development, Testing & Evaluation Working Group 39

Special Acknowledgements. 39

Appendix – Resourcing the Challenge 40

Definitions used in this document are taken from the NSTC Subcommittee on Biometrics' Biometrics Glossary where possible from http://www.biometrics.gov/Documents/Glossary.pdf.

About the National Science and Technology Council

The National Science and Technology Council (NSTC) was established by Executive Order on Nov. 23, 1993. This Cabinet-level Council is the principal means by which the executive branch coordinates science and technology policy across the diverse entities that make up the Federal research and development enterprise. Chaired by the president, the NSTC is made up of the vice president, the director of the Office of Science and Technology Policy, cabinet secretaries and agency heads with significant science and technology responsibilities and other White House officials. One of NSTC's primary objectives is to establish clear national goals for Federal science and technology investments in a broad array of areas spanning virtually all the mission areas of the executive branch. NSTC prepares research and development strategies that are coordinated across Federal agencies to form investment packages aimed at accomplishing multiple national goals.

NSTC established the Subcommittee on Biometrics in 2003 to develop and implement multi agency investment strategies that advance biometrics disciplines to meet public and private needs; coordinate biometrics-related activities of interagency importance; facilitate the inclusion of privacy-protecting principles in biometrics system design; ensure coordinated and consistent biometrics programs as government agencies interact with Congress, the press and the public; and strengthen international and public sector partnerships to foster the advancement of biometrics technologies. In 2007, the Subcommittee's charter expanded to address Identity Management, and it was renamed the Subcommittee on Biometrics and Identity Management (BIdM). The charter extends through December 2012.

About This Report

The NSTC Subcommittee on Biometrics prepared and published the original *National Biometrics Challenge* in August 2006. That report identified key challenges in advancing biometrics development. It was based upon analysis of the unique attributes of biometrics, the market forces and societal issues driving implementation of biometrics and the advances required for next-generation capabilities. A further prioritization was done within the Subcommittee, and the top third of priorities received about 83 percent of Federal funding.[1]

During the last five years, evolving mission needs, coupled with advances in technology, have necessitated a new look at research, development, test and evaluation (RDT&E) priorities. This 2011 update to the *Challenge* examines the many advances made as government, academia and the private sector responded to the "challenge" issued in 2006. It further delineates some of the complex issues that, five years later, have yet to be fully addressed. It acknowledges that the understanding of requirements has increased with experience while the advance of technology raises capabilities and expectations.

Government and industry have a common challenge to provide robust identity management tools and to deploy those tools intelligently to meet national and international needs. Collaboration among the biometrics community — government, industry and academia — on these common challenges remains essential.

---

1 The top third of priorities were those that were absolutely required to enable interagency interoperability, so they received a disproportionate amount of attention and resources. This funding ratio is not expected to continue into the future.

Executive Overview

America's national security community uses biometrics to resolve and then anchor the identity of known and suspected terrorists by linking information independently collected and maintained by the military, State Department Consular Affairs, Homeland Security's Customs and Border Protection (CBP) and Immigration and Customs Enforcement (ICE), state and local law enforcement, the Federal Bureau of Investigation (FBI) and other less widely known participants. Fingerprint and DNA forensic evidence, matched against an array of national biometric databases, allows identification and then capture of improvised explosive device (IED) makers. Aliens engaging in criminal activity are identified as such shortly after arrest, allowing ICE to target limited resources where they can have the greatest impact. Many thousands of persons, wanted for criminal offenses in distant jurisdictions, are identified each month while in custody for relatively minor offenses. The Nation is in the early stages of positively identifying suspicious persons at the point of encounter, separating the dangerous and actively wanted from the innocent; allowing the later to proceed upon their way with minimal disruption while focusing scarce resources where they can have the greatest impact. These and other applications for biometric technology, foreseen when the *Challenge* was issued in 2006, have become practical and operational reality. The Nation truly responded to the challenge.

A host of technical advances, policy changes, interoperability challenges, and organizational and cultural adjustments were required, achieved and overcome to bring these capabilities into being. The following report looks back upon the progress made during the past five years; it then looks forward at what remains to be done and at the opportunities available to strengthen national security and enhance public safety, but it also looks at opportunities to facilitate commerce through reducing the impact of identity theft and making the use of automated systems and networks more convenient.

The Nation has come a very long way in biometrics since 2006, yet high-quality capture devices still cost too much for many commercial applications. Even though face recognition technology has improved by nearly a factor of 10 over this period, further research is required to address applications where images are collected outside of a studio or mugshot environments. Rapid DNA analysis has advanced to the point it can soon be tested within the laboratory, and may in several years be introduced at the point of encounter, and take its place as a mature biometric for several national security and public safety applications. Again, more work is needed to get the cost of the rapid DNA technology on par with other biometric approaches and have the processing times accommodate application needs. Many other such needs and opportunities are addressed in this report.

In America's free society, there are also social, legal, privacy and policy considerations in government and commercial programs related to automated identification and identity management. While there is not yet a national consensus on these matters, there has been much progress in understanding the issues and progress towards such consensus. For example, the *National Strategy for Trusted Identities in Cyberspace* (NSTIC) is establishing the framework for use of trusted identities in cyber applications. In the *REAL ID Act of 2005*, Congress recognized significant shortcomings in the Nation's approach to identity documents as well as a

Biometrics
A general term used alternatively to describe a characteristic or a process.

As a characteristic:
A measurable biological (anatomical and physiological) and behavioral characteristic that can be used for automated recognition.

As a process:
Automated methods of recognizing an individual based on measurable biological (anatomical and physiological) and behavioral characteristics.

Recognition
A generic term used in the description of biometric systems (e.g., face recognition or iris recognition) relating to their fundamental function. The term "recognition" does not inherently imply verification, closed-set identification or open-set identification (watch list).

Verification
A task where the biometric system attempts to confirm an individual's claimed identity by comparing a submitted sample to one or more previously enrolled templates.

Identification
A task in which the biometric system searches a database for a reference matching a submitted biometric sample and, if found, returns a corresponding identity. A biometric is collected and compared to all the references in a database. Identification is "closed-set" if the person is known to exist in the database. In "open-set" identification, sometimes referred to as a "watch list," the person is not guaranteed to exist in the database. The system must determine whether the person is in the database, then return the identity.

---

public interest in increased identity protection. Ongoing dialogue and collaboration amongst all stakeholders to address these nontechnical matters remains essential.

With a few prominent successes as well as some continuing challenges in mind, it is timely to update and reissue the *Challenge*. The report examines the progress that has been made in response to the challenges laid out in 2006. Interagency interoperability and technology advancements in fingerprint, palmprint, face, iris, voice, multimodal biometrics and DNA are discussed. Various success stories are shared from Federal operations that have come online since 2006, including Secure Communities used by ICE in combination with state and local law enforcement agencies, mobile and maritime use of biometrics by the U.S. Coast Guard (USCG), and the Next Generation Identification (NGI) and face recognition programs supported by the FBI. Then the contemporary environment in 2011 is described including the technology environment, cost and institutional constraints, privacy expectations and issues of anonymity. A discussion of user-centric biometrics raises several key issues regarding possible directions over the next few years. The *Challenge* also discusses research challenges and focus areas for the science and technology community. With the increasing complexity of biometrics systems and issues, the continued need for education and training are emphasized, as are the accomplishments, trends and challenges surrounding standards, testing and certification.

The National Biometrics Challenge

1. Introduction

Government agencies have adopted biometrics for a variety of applications. For example, the criminal justice community, domestically and internationally, has been engaged in precursors to biometrics since the 1870s. In 1907, the Department of Justice (DOJ) established a Bureau of Criminal Identification, based upon fingerprints, and in 1924 charged the then-Bureau of Investigation with establishing a national identification and criminal history system that today is the Criminal Justice Information Services (CJIS) division of the FBI. CJIS operates the national criminal history and fingerprint based identification program using the Integrated Automated Fingerprint Identification System (IAFIS). IAFIS was deployed in 1999 as the culmination of 36 years of sustained research and development. At its peak, IAFIS contained fingerprint supported criminal history information on approximately 62 million persons, completing an average of about 168,000 fingerprint check requests daily, generating a response on average in a little over 6 minutes, and identifying more than 30,000 fugitives each month. More than a third of those identified are interstate fugitives who could not otherwise have been identified.

In addition, IAFIS also performs checks for disqualifying criminal history at the national level for employment in positions of public trust; positions dealing with vulnerable populations including children and the elderly, such as day care facilities, nursing homes and hospitals; positions in financial institutions; many activities requiring licenses; and other specified occupations for which the states and Congress have directed a noncriminal justice civil fingerprint supported check. IAFIS is at the end of its useful lifetime, routinely operating at more than double its designed capacity, and the government is transitioning to an NGI system that will be fully operational in the summer of 2014. NGI will deliver performance levels and support volumes of activity far beyond IAFIS.

While those in the 19th century addressed their identity management concerns in an appropriate way for their time, modern society is far more complex. Birth certificates, naturalization papers, passports and other government issued documents prove citizenship. To augment these well-established and familiar tokens, biometrics has emerged as a reasonable and effective way to identify individuals.

Other national biometric systems grew out of a desire to protect the borders and prevent terrorism. In the *Illegal Immigration Reform and Immigrant Responsibility Act* of 1996, Congress mandated the former Immigration and Naturalization Service (INS) to implement an automated entry and exit data system to track the arrival and departure of all aliens.[2] Today, the U.S. Visitor and Immigrant Status Indicator Technology (US-VISIT) Program, part of the Department of Homeland Security (DHS), operates the Automated Biometric Identification System (IDENT). IDENT maintains fingerprints, photographs and biographic information on more than 126 million individuals and conducts about 250,000 biometric transactions per day, averaging 10 seconds or less per transaction. The fingerprint scans and photographs captured for most non-U.S. citizens upon entry into the United States are stored in

Cooperative User An individual that willingly provides his or her biometric to the biometric system for capture. (Example: A worker submits his or her biometric to clock in and out of work.)

Non-cooperative User
An individual who is not aware that a biometric sample is being collected. (Example: A traveler passing through a security line at an airport is unaware that a camera is capturing his or her face image.)

Uncooperative User
An individual who actively tries to deny the capture of his or her biometric data. (Example: A detainee mutilates his or her finger upon capture to prevent the recognition of his or her identity.)

---

2 Congressional Research Service, *U.S. Visitor and Immigrant Status Indicator Technology (US-VISIT) Program* (Feb. 23, 2005, RL32234), by Lisa M. Serghetti and Stephen R. Vina.

Kandahar, Afghanistan

On April 25, 2011, the *New York Times* reported that more than 475 inmates, many of them Taliban insurgents, broke out of a prison on the edge of Kandahar in southern Afghanistan through an elaborate tunnel system. Within days, 35 of the men were recaptured by Afghan officials — thanks, in part, to biometrics. Fingerprinting, iris scans and facial recognition checks were conducted at border crossings, routine traffic stops and other key locations. One escapee was recaptured at a recruiting station while trying to join and infiltrate security forces. The men were rounded up thanks to "hits" based on earlier biometric enrollments as part of their prison in-processing. "You can present a fake identification card," SGM Robert Haemmerle, Combined Joint Interagency Task Force 435, told the *New York Times*. "You can shave your beard off, but you can't change your biometrics."

the system. IDENT checks have helped identify more than 110,000 wanted criminals, immigration violators, and known or suspected terrorists.

In response to the tragic events of Sept. 11, 2001, the Department of State (DOS) added a photograph repository and face recognition component to the Consular Consolidated Database (CCD) and began screening photos of visa applicants in 2004. CCD contains more than 110 million visa cases and nearly 90 million photographs and enables DOS to identify individuals who were previously denied or had their visas revoked or seeking multiple visas under different names. Visa applicants are cross-checked with IDENT, which is checked against IAFIS, for information that would disqualify them for admission to the United States.

In 2004, the Department of Defense (DOD) deployed the Automated Biometric Identification System (ABIS) to support military operations overseas. DOD built upon technology the FBI developed, adding palmprint, face and iris matching to establish a multimodal fusion capability unique in the Federal government. ABIS maintains fingerprint, photograph, iris and limited biographic information on about 6 million subjects and is increasing its transaction capacity from 8,000 to 20,000 per day.

The IAFIS, IDENT and CCD systems are three of the largest biometric identification systems in the world, while the more targeted and smaller ABIS is also a major biometric system. Collectively these systems represent a multi-billion dollar investment of the U.S. government and reflect the importance of biometrics and identity management at the Federal level.

2. Challenges Met and Remaining Since 2006

The 2006 *Challenge* set out many aggressive goals leading to significant accomplishments in a number of areas along with successes that have materially contributed to public safety and national security. Examples include:

- Advancing biometrics sensor technology for various modalities;
- Significant improvements in large scale systems performance;
- Allowing and promoting interoperability between systems;
- Establishing comprehensive and widely accepted open standards for biometric information, and the devices that capture it, to include conformance testing processes for broadly accepted certification;
- Protecting individual privacy and promoting public confidence in biometric technology and systems; and
- Developing a consistent and accurate message across the biometric community.

While none of these goals has been so completely satisfied that it can be said to be retired with no further work needed, very significant progress has been made over the past five years, particularly in interoperability and technology.

2.1 Interoperability

Advances in interoperability are an important success of the Federal biometric program over the five years since publication of the *Challenge*. When the *Challenge* was published in 2006, the interoperability of biometrically supported national security and public safety identification systems was a preeminent challenge. Some of the national identification systems, which needed to communicate to protect the Nation and its citizens, could not interact directly or in an automated way. Full interoperability between the FBI's IAFIS and DOD's ABIS was achieved in December 2005, and many battlefield detainees and other persons of interest encountered in combat zones were being routinely checked for previous U.S. domestic criminal justice activity. While there were understandable delays in transferring information collected on the battlefield to the national DOD ABIS, further linkage to the FBI IAFIS occurred in real-time.

Two-finger biometric matching of visa applicants, between the DOS CCD system and the DHS-operated IDENT system began in 2004, to ensure persons determined by DHS to be ineligible were not inadvertently granted visas. Otherwise, while statutes and national policy guidance were in place,[3] interoperability between national systems was limited and did not occur in real-time.

In September 2006, the FBI and DHS US-VISIT deployed an interim Data Sharing Model (iDSM), which provided a reciprocal exchange of biometric data subsets between IAFIS and IDENT in near real-time. This connection provided select

DHS Refugee Affairs Division

On May 27, 2011, DOD received a submission for an individual who applied for U.S. immigration benefits through the DHS Refugee Affairs Division. This resulted in a DOD ABIS match to a 2004 enrollment, when the individual had been detained for interfering with an investigation by stealing evidence. His DOD status was Tier 4 (Do Not Hire, Deny Base Access and Disqualify for Police or Army Training), and DHS screeners were able to take this additional information into account when adjudicating the man's immigration application. This match is just one example of the interoperability DOD and DHS have achieved between their respective databases and of the additional protection this collaboration provides to the Nation.

---

[3] Congress first mandated an automated entry and exit data system that would track the arrival and departure of every alien in the *Illegal Immigration Reform and Immigrant Responsibility Act* of 1996 (*IIRIRA*, P.L. 104-208). There are five principal laws that extend and refine *IIRIRA* to require development and implementation of an integrated entry and exit data system: The *INS Data Management Improvement Act* (*DMIA*; P.L. 106-215); *The Visa Waiver Permanent Program Act* (*VWPPA*; P.L. 106-396); *The Uniting and Strengthening America by Providing Appropriate Tools Required to Intercept and Obstruct Terrorism Act* (*USA PATRIOT Act*; P.L. 107-56); *The Enhanced Border Security and Visa Entry Reform Act* (*Border Security Act*; P.L. 107-173); and *The Intelligence Reform and Terrorism Prevention Act of 2004* (P.L. 108-458). Developing the systems and infrastructure to comply with this direction, and then linking systems across four departments of the Federal government was to prove a major effort.

Gulf of Aden

On May 15, 2009, U.S. Navy personnel from the USS *Gettysburg* sent biometric files to BIMA on 17 suspected pirates enrolled during an anti-piracy operation in the Gulf of Aden and documented their suspicious activity. By biometrically linking activities to individuals, legal processes can consider previous actions and prosecute individuals to the fullest extent of the law. Additionally, the increasing use of forward deployed biometric systems that share data and deny anonymity provides a continuous deterrent to pirate activity.

state and local law enforcement and authorized noncriminal justice agencies access to biometrically based immigration identity information for the first time. During the period of iDSM operation, 31,642 aliens were identified as wanted persons, and 158 were identified as known or suspected terrorists. Subsequently, in late October 2008, Shared Services was deployed, which provided interoperability participants access to the full IDENT repository. Today, state and local law enforcement agencies from jurisdictions in 43 states, one territory and five Federal agencies access IDENT information through IAFIS. Over the years, they have submitted nearly 12.9 million searches with about 8 percent resulting in matches. The search of the full IDENT repository has been extended to additional non-criminal justice agencies, along with mobile initiatives both domestically and internationally. Biometric interoperability has proven valuable to public safety, and DHS has been requested to authorize all domestic criminal justice agencies access to IDENT/IAFIS interoperability.

Through Shared Services, the FBI now shares additional data sets with DHS. When a search results in a biometric match to a DHS independent encounter, the information is retained in IDENT. In December 2007, IDENT began submitting CBP collected tenprint checks from air and seaports of entry primary processing lanes for full search of the FBI maintained national Criminal Master File (CMF). This resulted in the need for IAFIS to initially support 48,000 CBP searches per day. This has grown with IAFIS currently able to support up to 98,000 searches per day. The intent of this expanding effort is to enhance border security by identifying and denying aliens seeking admission to the United States who have prior U.S. domestic criminal histories. In May 2010, the FBI deployed a rapid response capability for use by the CBP in searches conducted against the FBI's CMF at ports-of-entry. In December 2010, DHS deployed this rapid response functionality to the Detroit International Airport. Since that time, the capability has expanded to three additional ports-of-entry with projections for full deployment to occur by January 2012. This functionality allows CBP primary officers to send individuals for secondary processing if the initial encounter results in a potential match to a CMF record.

In October 2006, the Homeland Security Council directed the DOS Consular Affairs and FBI to conduct a pilot program to assess the benefit of conducting fingerprint supported biometric checks against the CMF as part of the visa issuance process. Based upon the pilot results, which realized approximately a one percent hit rate, the decision was made to conduct such checks for all persons applying for visas worldwide. In January 2008, DOS began submitting about 30,000 tenprint checks per day, using IDENT/IAFIS interoperability. Half of these are high-priority transactions completed within 15 minutes. Through the end of July 2011, more than 88.5 million checks have been made by DHS, CBP and DOS. More than 826,000 resulted in positive identifications.

Positive steps have been taken to improve DOD's interoperability policies and capabilities, which represent a vital effort to complete the Federal biometrics "triad" comprised of establishing direct interoperability and connectivity between ABIS, IDENT and IAFIS. In September 2009, DOD and FBI signed a Memorandum of Understanding (MOU) allowing for deeper integration between ABIS and IAFIS, and in March 2011, DOD and DHS signed a MOU that provides the policy framework for interoperability between ABIS and IDENT. The flow of shared information between the agencies is, and will continue to be, multi-directional. As a demonstration of this interoperability, IAFIS connectivity with both ABIS and IDENT is currently being used by DHS ICE and the USCG to obtain a search and response of all three databases.

Many additional biometric interoperability successes have occurred in the past five years. A mobile initiative provides remote field biometric search access to national databases for appropriately authorized U.S. military and Federal law enforcement users. DOD has provided the FBI and DHS with digital copies of more than 150,000 latent fingerprints obtained from IEDs and weapons caches, as well as the biometric records of foreign individuals with ties to terrorist activity. The U.S. Office of Personnel Management, the DOS Office of Personnel Security and Suitability, and the FBI's Bioterrorism Risk Assessment Group is using IAFIS/IDENT interoperability to augment program results with biometrically supported checks on applicants for positions having public trust and/or national security implications.

Advances in interoperability are an important success of the Federal biometric program over the five years since the *Challenge's* publication. There is technical and policy work yet to be done, but much has been achieved. Tangible benefits have been realized from the progress. In addition to the recognized challenges that result from the proprietary nature of AFIS systems, there is also a need for improved interoperability for cross-jurisdictional latent fingerprint searches. Latent fingerprint interoperability is a crosscutting issue bringing together both the biometrics and forensics community.[4] An AFIS Interoperability Task Force has been established out of the NSTC Subcommittee on Forensic Science to establish a national strategy and roadmap, which will point to gaps requiring further research and solutions, such as limited funding and resources. A current effort to improve latent interoperability is underway as the FBI works with DHS and DOD to use established interoperability frameworks to pilot latent interoperability with select state and local law enforcement.

2.2 Technology

Biometric systems that assist Federal, state, local and tribal criminal justice practitioners with their duties are becoming more widely available, affordable and usable. Since 2006, progress across several modalities has been led by enhancements in technology, an expanded user base beyond the highly recognizable programs such as IAFIS, IDENT and ABIS, and the wide acceptance of e-passports and e-visas for automated border control operations. There has been increased implementation of open architectures for identification systems, with the adoption of Service Oriented Architectures (SOAs) using Web services by both US-VISIT and the FBI's NGI program. This has allowed the integration of multiple modalities within a single identification framework.

Basic and Applied Research

A vast array of basic and applied research have continued to take place in response to the 2006 *Challenge*. This research has not only enabled performance improvements and operational successes referred to in this *Challenge* document, but a strong research agenda and support are required to continue to substantiate the science and meet future operational gaps. Research since 2006 has been conducted to:

Tucson Sector Border Patrol Agents Prevent Sexual Predators From Entering U.S. (June 2011)

In four separate incidents, Tucson Sector Border Patrol agents arrested illegal aliens who were identified as sexual predators during processing.

On June 10, agents discovered that a 24-year-old Mexican national, apprehended near Naco, Ariz., had an active and extraditable warrant for aggravated criminal sexual abuse out of Woodstock, Ill. The man was turned over to the Tucson Police Department for extradition.

On June 11, Ajo Station agents arrested two sex offenders in separate incidents. The first subject was identified as a 34-year-old illegal alien from Mexico who was convicted in Fresno, Calif., of sexual assault for a lascivious act with a child under 14. He is being prosecuted for reentry of an aggravated felon. Also on June 11, agents apprehended a 24-year-old man from Oaxaca, Mexico, in possession of a mass data storage device containing child pornography. The man was turned over to ICE's Homeland Security Investigations for further investigation.

On June 12, Casa Grande agents arrested a 34-year-old Mexican national with a third-degree rape conviction in Oregon. The subject is being prosecuted for reentry of an aggravated felon. The criminal histories of all the subjects were exposed after their fingerprints were scanned into IAFIS.

Agents use IAFIS to access criminal records throughout the United States, and it rapidly identifies criminal aliens so they can be brought to an appropriate law enforcement resolution.

---

4 The National Academies, *Strengthening Forensic Science in the United States: A Path Forward.* In Recommendation 12, the report identified the need "to launch a new broad-based effort to achieve Nation-wide fingerprint data interoperability."

Afghanistan

On Nov. 16, 2010, USSOCOM detained and biometrically enrolled two individuals during a raid on a sensitive objective in Afghanistan. The men claimed to be innocent bystanders, and with no other derogatory information the team released them. A month later, USSOCOM encountered the same men on another raid at a different location. Although the Afghan men claimed to be someone else, they were quickly identified through their biometrics as being the same men from the first location. The men were detained for questioning, and their biometrics were added to the DOD watch list.

- Improve biometric modality performance and robustness;
- Develop new modalities;
- Optimize multimodal and large-scale fusion approaches and system designs;
- Conduct quality assessments, enhance quality and include quality in fusion approaches;
- Develop tools, statistical methods and modeling frameworks for system designs;
- Study socio-legal and business cases impacting biometric systems;
- Assess vulnerabilities in biometric devices and systems; and
- Fuse with results from related fields such as deception detection, credibility assessment, biographic data analysis and personnel disambiguation via 'soft' biometrics.

For more than 10 years, the National Science Foundation (NSF) has funded an Industry/University Cooperative Research Center (I/UCRC) for Identification Technology Research (CITeR) that has conducted more than 100 research projects.[5] The Center is composed of three NSF university research sites that work with numerous academic partner universities to conduct research that 19 corporate and Federal affiliates of the Center cooperatively sponsor and transition. The NSF I/UCRC program has a more than 30-year track record of developing industry-university centers organized around meeting the needs of their affiliates, and CITeR has seen strong growth in the past years based on this cooperative model.

In addition to CITeR, there are a number of strong university centers and labs focused on biometrics. The programs offered and research conducted across these institutions has resulted in key publications, research challenge progress and the training of students at all levels in biometrics. Graduates from these programs have gone on to directly support Federal and industry system development efforts.

Education

The biometrics community has provided strong support to Science, Technology, Engineering and Mathematics (STEM) education in support of achieving challenge objectives. Since 2006, a number of efforts have sought to provide the existing workforce with biometrics training. Educational institutions and organizations launched new offerings and increased the frequency of existing short courses to help meet immediate training needs. Delivery modes of these offerings include on-campus residential offerings, on-site offerings for specific clients and Web-based offerings. After extended development, the Institute of Electrical and Electronics Engineers (IEEE) established the Certified Biometric Professional program in 2010 to create a designation to help identify individuals who possess the knowledge and skill set essential for biometrics professionals.[6]

Individuals who pass the IEEE Certified Biometrics Professional examination are expected to have the proficiency needed to perform effectively in biometrics

---

[5] Center for Identification Technology Research (CITeR), last modified Aug. 24, 2011, http://www.citer.wvu.edu/.
[6] IEEE Certified Biometrics Professional Program, http://www.ieeebiometricscertification.org/.

programs. To facilitate individuals interested in earning Certified Biometrics Professional status, IEEE has developed a Learning System based on input from biometrics experts in industry, government and academia. This system is a comprehensive professional development and exam preparation tool. More than 100 individuals have passed the rigorous examination and agreed to follow the Certified Biometrics Professional code of ethics.

At the associate and bachelor levels, institutions have increased their offerings that address biometrics primarily in the context of forensics or security-based programs, emphasizing areas and courses that address law enforcement or identity management applications. There is one U.S. engineering-based B.S. program in biometric systems, and many institutions offer either specific courses or an emphasis in biometrics that may be earned in conjunction with a degree program.

Numerous U.S. universities are among the many international schools and institutes that provide doctoral and master of science students training in biometrics and related fields through dissertation and thesis research. Typically, these students major in very specific technical disciplines, such as computer and/or information science, computer engineering or electrical engineering. Professional masters programs provide course work-based training, typically in the context of forensics or information security.

Fingerprint and Palmprint

Fingerprints are arguably the most widely recognized biometric, particularly for criminal justice applications, and in support of border security and identity proofing. With well-known and established large-scale AFIS systems in place to manage fingerprint records and searches, friction ridge recognition technology has advanced in a number of significant ways since 2006. Field collection methods have improved, including the development of mobile capture devices for point-of-encounter identification. This advancement is significant, and it underscores the need for capture devices to continue to mature to meet the requirements for submission to enhanced systems (e.g., a livescan device that evolves to allow for the collection of supplemental fingerprints). Improved accuracy rates for tenprint searches reached 99 percent through algorithm enhancements. Advanced analytical tools and image processing technologies specialized for latent fingerprints, such as a latent background noise removal algorithm and a low-quality fingerprint ridge recognition algorithm, have resulted in significantly improved accuracy rates for latent matching. Expanded law enforcement use of palmprint matching using friction ridge recognition technology is another important advancement.

Recent deployments of fingerprint quality assessment algorithms (including the National Institute of Standards and Technology (NIST) Fingerprint Image Quality (NFIQ)) in large-scale identity management applications has resulted in the need to update the technology as well. The development of an open source NFIQ 2.0 is currently underway.[7]

The FBI has followed suit with the extension of its Appendix F standard to include 1,000 pixels per inch (PPI) images and the introduction of a new program to certify devices intended for use in the Federal Information Processing Standard Publication 201 (FIPS 201) Personal Identity Verification (PIV) program. This new

---

Tucson Sector Border Patrol Agents Arrest Five Illegal Aliens with Criminal Histories (June 2011)

Tucson Sector Border Patrol agents arrested five illegal aliens with dangerous criminal histories. Three of the illegal aliens are members of violent street gangs, and two had felony convictions for manslaughter.

Casa Grande Station agents patrolling near Sells, Ariz., apprehended an illegal alien from Mexico who was identified as a member of the Mara Salvatrucha 13 street gang. Later that evening, agents from the Ajo Station apprehended an illegal alien from Mexico northeast of Lukeville, Ariz., who admitted to being a member of the Sureños.

Naco Station agents apprehended an illegal alien from Mexico who admitted to being affiliated with the 18th Street gang. All three subjects are being criminally prosecuted for illegal entry. On Friday, an illegal alien from Tamaulipas, Mexico, was apprehended near Amado, Ariz., by Nogales Station agents. Record checks revealed the suspect had a prior conviction in Queens, N.Y., for first-degree manslaughter with intent to cause serious physical injury.

Ajo Station agents patrolling northeast of Lukeville apprehended a Mexican national with an extensive criminal history including convictions in California for voluntary manslaughter, robbery, burglary and assault with a deadly weapon. Both subjects are being prosecuted for reentry of an aggravated felon.

---

[7] NIST Image Group Fingerprint Overview, last modified April 14, 2011, http://www.nist.gov/itl/iad/ig/fingerprint.cfm.

standard, the PIV-071006 is a lower-level standard designed to support one-to-one (1:1) fingerprint verification scanners and is in line to support expanded applications of biometrics in PIV cards.[8]

Face

Biometric face recognition has seen a number of advances since 2006, driven by the trend and popularity of social networking sites, the prevalence of mobile smartphone applications and successful implementation in visa applications and in criminal and military investigations. Media giants such as Google, Apple and Facebook now include face recognition in their products, and the commercial development of low-cost "face sensors" (cameras with built in face detection) is underway. The main challenge for face recognition, as with any non-contact biometric, continues to be compensation for unconstrained individuals and environments and the use of low quality sensors. Face recognition, for example, on a low resolution video image taken outdoors with harsh shadows continues to be a challenge; although a high resolution studio, credential or booking photograph taken with controlled pose, lighting and background now performs well for many applications. The measured error rate in face recognition has continued to drop by half every two years.[9] Other advances in face recognition include recognizing faces in a crowd and three-dimensional face recognition.

To improve face recognition performance in controlled applications, the ANSI/NIST-ITL 1-2000 standard, *Data Format for the Interchange of Fingerprint, Facial and Scar, Mark and Tattoo (SMT) Information* was replaced in 2007 by an updated ANSI/NIST-ITL 1-2007 that includes best practice application levels for the capture of face images. By conforming to the standard, law enforcement, criminal justice agencies and other organizations that process face data can exchange facial or other photographic images and related biometric identification data.[10]

Beyond basic image-to-image comparison of faces, there have also been breakthroughs in 'video-to-video' matching and 'still-face-to-video' matching. Improvements in face recognition technology will continue with programmatically focused investments by government agencies. For example, the U.S. Special Operations Command (USSOCOM) funded a proof-of-concept effort to create handheld, mobile binoculars capable of automatic face recognition at ranges up to 100 meters in outside daylight. The National Institute of Justice (NIJ) provided additional development support to advance the proof-of-concept prototype for controlled testing and evaluation. Efforts like this contribute to improved understanding of how biometrics can best be used for operational applications and provide tools to address known technological gaps.

Iris

Iris recognition has shown significant improvement since 2006 in both capture devices and recognition algorithms.[11] The number of camera manufacturers and models available has doubled from a decade ago. These newer cameras offer much lower failure-to-capture rates and transaction times, and some have the ability to collect

---

NIST – National Institute of Standards and Technology

A non-regulatory Federal agency within the U.S. Department of Commerce that develops and promotes measurement, standards and technology to facilitate trade and improve the quality of life. NIST's measurement and standards work promotes the well-being of the Nation and helps improve, among other things, the Nation's homeland security. For more information, visit www.nist.gov.

ANSI – American National Standards Institute

A private, nonprofit organization that administers and coordinates the U.S. voluntary standardization and conformity assessment system. ANSI's mission is to enhance both the global competitiveness of U.S. business and quality of life by promoting and facilitating voluntary consensus standards and conformity assessment systems and safeguarding the standards and systems' integrity. For more information, visit www.ansi.org.

ISO – International Organization for Standardization

A nongovernmental network of the national standards institutes from 151 countries. The ISO acts as a bridging organization in which a consensus can be reached on solutions that meet both the requirements of business and the broader needs of society, such as the needs of stakeholder groups like consumers and users. For more information, visit www.iso.org.

---

8 The FBI IAFIS Certified Product List is available from https://www.fbibiospecs.org/IAFIS/Default.aspx
9 NIST Face Homepage, last modified Feb. 15, 2011, http://www.nist.gov/itl/iad/ig/face.cfm
10 An update to the ANSI/NIST ITL Standard is underway, and it is expected to be adopted by the end of 2011. NIST ANSI Standard Homepage, last modified July 19, 2011, http://www.nist.gov/itl/iad/ig/ansi_standard.cfm
11 NIST Iris Homepage, last modified May 24, 2011, http://www.nist.gov/itl/iad/ig/iris.cfm

iris images at a distance and in motion. Additionally, there has been a migration from single-eye to two-eye capture, constituting a more powerful biometric by removing left-right ambiguity and providing faster one-to-many (1:many) operations.

With the expiration of some iris patents circa 2005, there are now many more iris recognition algorithm manufacturers. The number of providers has increased to more than 10, greatly expanding availability. These newer algorithms operate on rectilinear images formally standardized according to ISO/IEC 19794-6:2011 and ANSI/NIST ITL 1-2011. The standards support general formats and compact mode, a few kilobytes for 1:1 operation, and a few tens of kilobytes for 1:many operations. This new generation of iris cameras also offers on-board storage for iris templates, encryption capabilities and software for host-based matching of larger template databases. This is particularly useful for field applications that face the challenge of limited data transfer capabilities. The use of iris recognition applications in large-scale identity programs, such as the Unique Identification Authority of India (UIDAI) project and in corrections facilities to eliminate errors related to mistaken identity, have propelled advances in this biometric modality.

To support the expanding iris marketplace and ensure that iris images and templates can be shared, three Iris Exchange (IREX) activities were initiated.[12] The first, IREX I, addressed standards, formats and compression for data interchange. The second, IREX II, is intended to define and measure image quality. The third, IREX III, will give guidelines to users to support large-scale identification applications.

Mobile Multimodal Biometrics

Numerous handheld mobile devices with the ability to collect and verify fingerprint, face and/or iris have also been demonstrated since 2006. Many of these multi-biometric devices now come with wireless communications capability, weigh just over one pound and have storage capacity for tens of thousands of records. These devices in general have dropped in cost, resulting in savings of $1,000 to $5,000 per device. Additional improvements include increased battery life and expanded memory capacity. As recently as April 2011, a multi-functional biometric access control reader was introduced that interfaces with any Physical Access Control System and can support government credentials such as PIV, Transportation Worker Identification Credential and the Common Access Card.

Very small, smartphone-based mobile devices with biometric capabilities are also in early development at this point. The familiar form factor of a handheld device is leading to increased acceptance by law enforcement for suspect identification in the field (early adoption at this point) and is gaining acceptance in commercial industries for applications such as accessing secure medical records and securing credit card transactions. The new mobile biometric devices allow first responders, police, military and criminal justice organizations to collect biometric data with a handheld device on a street corner or in a remote area and then wirelessly send it for comparison to other samples on watch lists and databases in near real-time. Identities can be determined quickly without having to take a subject to a central facility to collect his or her biometrics, which is not always practical.

FBI Certification Program

The program provides assurance to users of biometric collection systems that certified products meet or exceed minimum FBI interoperability standards and will work with IAFIS. These standards ensure images used are high quality and support all phases of identification for fingerprint experts and IAFIS. Standards used to certify products on the FBI Certified Product List are:

- **Appendix F** has stringent image quality conditions, focusing on the human fingerprint comparison and facilitating large-scale machine one-to-many matching operation.

- **PIV-071006** is a lower-level standard designed to support 1:1 fingerprint verification. Certification is available for devices intended for use in the FIPS 201 PIV program.

One-to-Many (1:Many)

A phrase used in the biometric community to describe a system that compares one reference to many enrolled references to make a decision. The phrase typically refers to the identification or watch list tasks.

One-to-One (1:1)

A phrase used in the biometrics community to describe a system that compares one reference to one enrolled reference to make a decision. The phrase typically refers to the verification tasks (though not all verification tasks are truly 1:1), and the identification task can be accomplished by a series of 1:1 comparisons.

---

12 NIST Iris Exchange (IREX) Activities, last updated May 5, 2011, http://www.nist.gov/itl/iad/ig/irex.cfm.

PPI

Pixels per inch is a measure of the resolution of a digital image. The higher the PPI, the more information is included in the image and the larger the file size.

Prince George's County, Md.

A subject was arrested during a traffic stop. Although he had used at least three aliases during previous encounters with law enforcement, his fingerprints check revealed an active FBI warrant related to a bank robbery investigation and illegal entry to the United States. FBI is coordinating with ICE to ensure that the subject is transferred into ICE custody upon disposition of the criminal investigation.

In order to ensure that biometrics collected with mobile devices work with traditional stationary systems, *NIST Special Publication 500-280: Mobile ID Device Best Practice Recommendation Version 1* was developed.[13] The FBI has adopted these mobile best practices for fingerprints into its Electronic Biometric Transmission Specification (EBTS) and subsequently administers an image quality certification program for the various profiles of mobile device capture capabilities as a service to the U.S. government.

Voice

Applications of biometrics for secure access have brought increased attention to the use of speaker recognition. Since 2006, speaker verification (1:1 matching) and speaker identification (1:many matching) have expanded significantly for both commercial and noncommercial applications, with improving performance.[14] The advances in speaker recognition technology are attributed to many factors including:

- Advanced algorithms to deal with cross-channel effects and speaker variants;

- Increased computing power;

- Fast query and weighing algorithms that enable the processing and fusing of output from multiple algorithms and models; and

- New hardware devices specialized for clear capture of audio input while cancelling ambient background noise.

In spite of the advances made in speaker verification since 2006, speaker recognition as a biometric application is still faced with many challenges including the lack of a standard measurement for acceptance, difficulties with the capture of a consistent voice sample at enrollment and the performance issues associated with the lack of comparable recording environments between the enrollment and test sample.

DNA

DNA analysis is universally acknowledged as a significant tool for fighting crime and is the only biometric that contains specific characteristics passed from parent to child. DNA is currently used to perform forensic identifications and/or provide leads in criminal, missing person and terrorism investigations. Recent technological advances are now being employed in the development of portable Rapid DNA machines that are being designed for use by law enforcement officers in booking stations to initiate DNA analysis of arrested individuals much more expeditiously than in the past. In addition, this machine has applications for soldiers in theater to identify detainees and for immigration and border agents to confirm individual identifications or family relationship claims. This "swab-in/profile-out" capability is poised to provide a new tool for rapid identification outside of the forensic laboratory where DNA technology is traditionally applied.

---

13 *NIST Special Publication 500-280: Mobile ID Device Best Practice Recommendation Version 1*.
14 The missed detection error rate in speaker verification has dropped from 16 percent in 2006 to approximately 7 percent in 2010 when utilizing telephone channels and holding the false alarm rate constant at 10 percent. *NIST Speaker Recognition Evaluation*, last modified March 23, 2011, http://www.nist.gov/itl/iad/mig/sre.cfm.

Although not yet ready for field or laboratory use, Rapid DNA is not science fiction. With a joint prototype development program by NSTC participants, and several government agencies incentivizing development efforts, first generation Rapid DNA prototype systems are expected to soon be available for evaluation. NIST has conducted research on the stability of human DNA collected on various swab materials and are developing testing plans for these prototype devices. This evolution in using DNA for point-of-collection analysis can be seen as a significant advancement in the use of biometrics for criminal, immigration and counterterrorism applications.

2.3 Notable Successes

The previous sections on interoperability and technology present a partial catalog of technical successes over the past five years. However, the real impact upon society has come from the operational employment of the new technical capabilities coupled with policies, procedures, processes and the institutional resources of the criminal justice and national security communities.

Through Secure Communities, ICE improves public safety every day by transforming how criminal aliens are identified and removed from the United States. This strategy leverages information sharing capability between DHS and the DOJ to identify aliens who are arrested for a crime and booked into local law enforcement custody quickly and accurately. With this capability, the fingerprints of everyone arrested and booked are not only checked against FBI criminal history records, but also against DHS immigration records. If fingerprints match DHS records, ICE determines if immigration enforcement action is required, taking into consideration the immigration status of the alien, the severity of the crime and the alien's criminal history. Secure Communities also helps ICE maximize and prioritize its resources to ensure that the right people, processes and infrastructure are in place to accommodate the increased number of criminal aliens being identified and removed. Since this capability was first activated in 2008, biometric information sharing has helped ICE identify more than 150,000 convicted criminal aliens that were administratively arrested or booked into ICE custody. More than 77,000 convicted criminal aliens have been removed from the United States.

Beginning in early 2007, DOD expanded the use of biometric technology from its traditional role in military law enforcement, enabling biometrics' usability to cut across various realms of use. Thousands of portable, handheld biometric verification devices were issued to Soldiers on patrol in Afghanistan and Iraq to check the biometrics of suspicious individuals against watch lists developed from the latent fingerprints found on improvised explosive devices or weapons caches. These biometric devices assisted local leaders in establishing resident lists so that insurgents could not be assigned positions of trust or allowed entry into secure locations, developing employee rosters to prevent attacks on critical infrastructure sites, and identifying insurgents attempting to infiltrate local security forces. Similar devices were deployed to most Forward Operating Bases in Afghanistan and Iraq to check local workers against these watch lists, and ensure that known insurgents could not slip onto base by showing forged papers, and the data from all systems was collected and matched at the ABIS database.

As a result, DOD has biometrically identified more than 150,000 non-U.S. persons with suspicious records and more than 160,000 latent fingerprints. These

---

Catawba County, N.C.

Three subjects were arrested and booked on charges of first-degree rape, first-degree kidnapping and robbery with a dangerous weapon. Their fingerprints check revealed they were in the country illegally. ICE placed detainers for all three subjects and will remove them from the country upon completion of their criminal sentences.

New Castle County, Del.

A subject was arrested for possession of marijuana and drug paraphernalia. His fingerprint check revealed that he had used an alias upon booking, had been previously deported and had a lengthy criminal history in three states. This included convictions for burglary, carrying a concealed weapon, drug-related crimes and charges for gang affiliation. He had previously tried to enter the United States illegally three times using aliases and claiming citizenship from another country. ICE is detaining the subject pending removal.

Florence, Ariz.

A man was identified as a recidivist aggravated felon after local police arrested him for providing false information to law enforcement. He had been deported nine times and had a criminal history of 51 arrests under 16 aliases. ICE placed a detainer for the subject pending his disposition and will remove him from the United States.

identifications are shared between agencies for homeland security screening, and there have been a significant number of cases where individuals were refused entry into the US using this shared information.

Based on these and other successes, the DOD's use of biometrics has grown significantly. DOD employs biometrics to allow warfighters to accomplish operational and institutional missions across the spectrum of operations. These missions can include force protection, security and stability operations, personnel recovery, disaster relief, humanitarian assistance, high-value individual identification and tracking, and information sharing for U.S. border protection and law enforcement.

The USCG is using mobile biometric technology, capturing fingerprints and photographs, to identify illegal migrants who are apprehended while attempting to enter the United States through the Mona Passage between Puerto Rico and the Dominican Republic and the Florida Straits. Under the program, the Coast Guard digitally collects fingerprints and photographs from illegal migrants apprehended at sea, then it uses satellite technology to transmit the biometric data and immediately compare the migrants' information against US-VISIT databases (IDENT) which, using interoperability with IAFIS, includes information about wanted criminals, known or suspected terrorists, immigration violators and those who have previously encountered government authorities. The program has significantly advanced the effort to develop effective mobile solutions for biometric collection and analysis. It represents another step in DHS' comprehensive strategy to secure the Nation's borders. Since the program began in November 2006, it has collected biometric data from 2,828 migrants, had 711 matches against the US VISIT IDENT database and supported prosecution of 362 of those migrants. Biometrics is one of the tools being used to identify and prosecute human traffickers and to discourage others from making the perilous crossing. As a result of these efforts, migrant interdictions are down 93 percent from Fiscal Year 2006 to today in the Mona Pass.

On Feb. 25, 2011, the FBI achieved initial operating capability for the NGI system, deploying advanced matching algorithms, which identified 910 additional candidates during the first five days of operation that the legacy IAFIS system missed. NGI fingerprint search reliability of 99 percent, when compared against a repository of 100 million persons, where NGI matched individuals who had been missed by IAFIS to existing criminal records, is a significant advance upon IAFIS' 95 percent search reliability. During the operational validation period, numerous identification successes gave evidence of improved performance, including identification of subjects with criminal charges of: first-degree murder, manslaughter, assault on a police officer, fugitive from justice, harboring aliens, smuggling aliens and felony abduction. In addition, during civil background investigations, NGI identified subjects with prior charges including a caregiver applicant charged with battery, an immigration applicant charged with rape and an immigration applicant charged with burglary.

In early 2008, the FBI launched the "Face Mask" initiative in collaboration with the North Carolina Division of Motor Vehicles (DMV). Face Mask uses face recognition software to apprehend fugitives and locate missing persons. The DMV already used a face recognition system to compare new driver license applicants with their database of 30 million driver license images, to identify individuals with revoked and suspended driver licenses and those attempting to obtain additional licenses fraudulently under assumed names. The FBI and DMV conducted a pilot project, performing searches of individuals wanted in North Carolina and the five surrounding states that led to more than 700 investigative leads including one Federal fugitive apprehension, six state fugitive apprehensions and one missing person resolution. Most notable was the identification and subsequent arrest and conviction of a fugitive wanted for murder.

The FBI's Biometric Center of Excellence engineered breakthrough technology, formally called Automated Face Detection and Recognition (AFDAR), for analysis and correlation of face images. AFDAR, also known as Cluster Base, is a forensic image analysis tool that locates faces within images and clusters them based on similarity. Complementary tools AFDAR-C and AFDAR-V process stills and video, allowing investigative agencies to analyze large collections of images and video recordings, conduct correlation across disparate sets of data and cluster images for link analysis and second level review. Cluster Base was successfully used in the 2010 "Golden Broker" case where an Asian criminal enterprise obtained names and Social Security Numbers from temporary workers, selling them as a base for fraudulent identification to in turn obtain passports, mortgages and credit cards. Investigators identified several states in which fraudulent identification was being used. Cluster Base searched more than 14,000 images to find offenders. Thus far, 53 New Jersey subjects have been charged with multiple offenses.

Zapata County, Texas

A subject was arrested for capital murder of his 6-month-old nephew. Upon being booked into law enforcement custody, his fingerprints were checked against DOJ and DHS records. ICE was immediately notified that the individual in custody was illegally present in the United States. Upon disposition of his current charges, ICE will assume custody and remove him from the United States.

Hillsborough County, Fla.

A subject was arrested for carrying a concealed weapon, opposing an officer and providing a false name to law enforcement. Despite his past use of multiple aliases, cross-checking his fingerprints revealed that he entered the United States legally as a B-2 non-immigrant visitor but did not leave when his visa expired. The subject also had an active criminal arrest warrant for attempting to murder a police officer. He was convicted for carrying a concealed firearm and sentenced to one year in prison. Following the completion of his sentence, ICE will assume custody and remove him from the United States.

3. The 2011 Environment

The biometric environment is characterized by the capabilities and technology of the current installations, on-going research, emerging technologies, institutional constraints, and privacy, civil rights, and civil liberty issues. Great progress has been achieved in national security, homeland security and law enforcement, however, additional emphasis is needed in e-government and e-commerce. Commercial applications are driven by value to the customer, cost and the inexorable march of consumer electronics towards increased mobility, increased computational capabilities, faster response and improved usage. However, factors slowing acceptance in the commercial arena and by potential users are existing institutional constraints and uncertainties associated with the use of biometrics. Many of these constraints are based on legacy practices; overcoming them will require providing accurate and sufficient information to the potential entrants into the larger commercial market place. Privacy, civil rights and civil liberties are fundamental and highly complex issues that also need to be addressed as part of the entry process.

3.1 Technology Landscape

As in 2006, the primary users of biometric technology are large government identification systems used by law enforcement, national security, military and border control (immigration management). There is also an expanding biometric credential program for identity verification and electronic access. The use of biometrics is gaining acceptance as a security measure at airports and other critical access sites in the United States and internationally. Interoperability between international systems and the U.S. identification systems is expanding in areas of national security and e-passport projects for U.S. visa waiver countries, EURODAC,[15] the Visa Information System and the new generation Schengen Information System. These systems are emerging as major drivers of biometrics usage.

Progress in government biometric applications has been significant. Major accomplishments can be summarized as follows:

- All national biometric systems have improved their capability to process very large workloads and accommodate increased database sizes while also improving accuracy and response times.

- The use of commodity hardware and SOAs has led to more flexible system architectures, which have facilitated technology improvement and the introduction of new capabilities. Using dedicated hardware to process biometric information is no longer widely practiced.

- A significant expansion of identification capabilities has been realized through greater interoperability between Federal agencies and their international partners. Data relating to potentially dangerous individuals can be exchanged consistent with established policies and standards, and data can be processed by all national identification systems to maximize the likelihood of identification and to provide linkage to data captured under a different identity.

- Standards have matured significantly and have contributed to improved system and biometric device interoperability.

San Pablo, Calif.

One day after activating the Secure Communities capability, a subject was arrested for vehicular manslaughter after he struck and killed a motorcyclist returning from Bible study. He was also charged with hit and run, driving under the influence (DUI) and driving with a suspended license. Cross-checking his fingerprints revealed his illegal immigration status, a prior removal and previous convictions for carrying a concealed weapon, cruelty toward spouse and multiple DUIs. ICE agents were automatically notified and placed a detainer for the subject. Upon conclusion of his judicial proceedings, the case will be presented to the Assistant U.S. Attorney's Office for prosecution for reentering the country after deportation. ICE will remove him from the United States once the justice process is complete.

---

15 EURODAC stands for European Dactyloscopy. It is the European fingerprint database for identifying asylum seekers and irregular border-crossers.

- In-theater, military operations have clearly demonstrated the value of multimodal biometrics and their benefits when used in highly diverse and difficult environments. These successes have increased the adoption of multiple modalities, which increase the likelihood of identification.

- The U.S. government has supported technology testing and standards development. This support has created a framework and a strong stimulus for continued technological improvement through coordinated and focused research and product development.

- The development and wide acceptance of lower cost biometric handheld devices now make it possible to obtain rapid identification virtually anywhere. The capability to make these rapid identifications is supported by the ability of national biometric systems to match biometric data.

- The impact of sample quality on biometric performance is now understood and usable performance metrics have been established.

- Improvements in communications have facilitated and will continue to improve the ability to exchange large volumes of image data that is at the core of most biometric processes.

These accomplishments have greatly strengthened and improved the ability of the national security, homeland security and law enforcement communities to accurately recognize individuals and distinguish between those who do and do not pose a threat in a wide range of operating conditions. However, improvements are still needed, and new needs have been identified. The systems of the future need to accommodate the following high-level improvements:

- Developing modular, plug-and-play software capable of accommodating ongoing growth in the amount and types (modalities) of collected biometric data. The increased proliferation of diverse biometric capture devices, even within the same modality, poses a challenge in obtaining the right algorithmic approaches and ensuring that the system software can be adapted to the quickly changing characteristics of the data capture devices and integrated with appropriately upgraded algorithms. This requires the development of comprehensive architecture, standards and testing frameworks capable of successfully exploiting the diverse data and technological improvement. In addition, methods need to be developed for effectively exploiting data to optimize identification accuracy under diverse operational requirements.

- Increasing industry competition in iris recognition technology has resulted in rapid growth in iris recognition capabilities and in lower costs. However, the potential benefits of this technology have not yet been fully realized on a national scale. Effective concepts of operation and methods for forensic analysis need to be developed.

- Algorithmic improvements in face recognition have been dramatic. While face recognition is being incorporated into the large identification systems, its potential is still not fully realized due, at least in part, to challenges in addressing the many variables in collection and matching (e.g., pose, illumination, expression and aging). Use of facial recognition as a possible

Hartsfield-Jackson International Airport

In August 2010, Customs and Border Protection officers at Atlanta's Hartsfield-Jackson International Airport arrested a man after fingerprint records confirmed outstanding warrants for his arrest in connection to a murder in Michigan. Interoperability between FBI and US-VISIT systems helped the officers obtain the information they needed in a timely fashion. The man, suspected of several murders and assaults in Michigan and Virginia, was arrested as he attempted to board a flight bound for Tel Aviv, Israel. The officers took the man into custody and turned him over to law enforcement authorities.

NGI's Repository for Individuals of Special Concern (RISC)

Proof of concept for NGI's RISC, a new capability to identify persons wanted on the most serious charges as well as known terrorists within 10 seconds, came from a citizen tip, which alerted a tactical unit a murder suspect was at a local shopping center. The subject gave a false name and date of birth. Officers fingerprinted the subject using the Mobile ID device with the RISC pilot, determined there was an active warrant for the murder. The suspect also had active warrants for attempted murder, attempted robbery and kidnapping.

Bureau of Consular Affairs

On July 30, 2010, 2:46 p.m., IAFIS received an electronic fingerprint submission from the U.S. DOS Bureau of Consular Affairs/Visa Office in Washington, D.C. The fingerprints were processed on IAFIS and hit against a pre-1960 record. While the fingerprints were included in the automated system, the associated biographical and criminal history information had not been automated.

A search of the manual files revealed the subject had one previous arrest for an immigration-related offense and one for petty theft. At 3:17 p.m., a response containing this information was returned to the submitting and wanting agencies.

US-VISIT

In November 2010, US-VISIT assisted in a case to determine the true identity and overstay status of a Turkish man attempting to gain employment at a nuclear power plant. It was determined that the subject was using a false document under a false identity to prove his legal status to reside and work in the United States. The subject was subsequently arrested by local DHS law enforcement authorities as a visa overstay and placed into Federal custody awaiting removal proceedings.

means of detecting identification fraud in combination with other biometrics is one application that could further be developed.

- Improving system tolerance for non-ideal biometric presentation and acquisition is needed to increase the ease of use, acceptance and throughput for cooperative users of civil and commercial systems. It is also needed to deal with non-cooperative users in law enforcement, defense and security deployments. This includes developing methods to effectively exploit biometric data to optimize identification accuracy under diverse operational requirements.

- Incidents of identity avoidance and spoofing have increased. While progress has been made in developing detection algorithms and in providing detection at the data-capture level, effective countermeasures need to be further implemented at a system level.

- The development of Rapid DNA systems has significantly advanced molecular biometrics. DNA processing systems capable of providing usable results for non-ideal and degraded samples, as well as systems for new classes of bio-molecular targets such as scent volatiles and microbial colonies, need to be developed.

- Growing world political instability and natural disasters have resulted in often unexpected and rapid large-scale population dislocations. This requires that systems collect and process large amounts of data quickly and accurately to determine benefit eligibility (humanitarian assistance) and simultaneously detect potential terrorists within the dislocated population.

- Increasing use of handheld biometric devices by the law enforcement and national security communities will require additional capacity growth and accuracy improvements to accommodate the increased computational workloads that these devices impose on the large identification systems. To minimize the additional workload on these large identification systems and facilitate interoperability, it is necessary to optimize mobile devices, especially regarding sample quality and usability.

Use of biometric systems has increased in the areas of private enterprise, e-government services personal information and business transactions. However, overall progress has been limited in scope. Among the most significant achievements is the GSA's FIPS 201 program, which has expanded the use of biometrically enabled credentials by many Federal agencies. However, their potential has not been fully realized. Primarily, the credentials are used for facility access control, although frequently a credential reader is not available, and the credential becomes nothing more than an expensive flash card pass. Credential use for logical access is also limited. Some issues relating to the wider use of the biometrically enabled credentials are reader cost, lack of an enterprise approach for identity management, poor usability characteristics of the local biometric credential authentication systems and lower than desired performance levels for identifying the credential holder. There are currently no published plans or frameworks for large-scale use of biometrics for logical access, as might be necessary for large-scale benefit systems and extensive e-government applications.

In commercial applications dealing with personal information and business transactions, other countries have made considerable progress, particularly Japan and South Korea, where biometrics have been extensively adopted by the banking industry (vascular pattern recognition). Some European countries are also beginning to adapt biometrics in the commercial arena. New smartphones with biometric capabilities are likely to have a large impact on how biometrics are used and on the biometric systems that support them. Many view these small, relatively inexpensive and compact devices as representing a breakthrough technology for the e-commerce market. Smartphone applications, which would be biometrically secured, are expected to include financial and commercial transactions as well as providing many forms of travel documents such as tickets and boarding passes.

A very large biometric project under way in India is the UIDAI, which plans to capture multimodal biometrics for a large portion of India's population of 200 million. A large portion of the general population of India has no other form of identification and biometrics will allow for access to all government programs and business transactions. If successful, the project may stimulate the production of inexpensive readers and point-of-sale terminals on a scale that is likely to result in cost reductions and in the development of a large biometric identification infrastructure.

3.2 Cost and Institutional Constraints

There has been slow progress in the U.S. in adopting biometrics in the commercial arena, outside of a few notable exceptions, such as Disney World. The reluctance to adopt biometrics appears to be due to a combination of factors such as cost, institutional factors, authentication security concerns and privacy concerns.[16] The primary cost and institutional factors that inhibit widespread adoption of biometrics are as follows:

- Many potential commercial users view the investment cost of implementing a biometrically based identity proofing system as excessive.

- Those enterprises that might implement an identity proofing system are likely to want a control of the credential holders' data.

- The commercial user cannot determine the cost/benefit tradeoff on replacement of legacy security protection mechanisms by a system relying on biometrics.

Adoption of biometric identity authentication requires development and integration of an identity services framework similar to the one implemented by the General Services Administration (GSA) as part of the FIPS 201 program. Many of the same elements are required:

- A process for establishing an identity for the potential customers and tying it to a secure credential;

- Implementation of biometric capture devices at all point-of-sale terminals; and

- A framework for processing the credential, authenticating the identity and tying it to the business transaction.

Chicago

On March 10, 2011, at 12:50 a.m., IAFIS received an electronic fingerprint submission from the Police Department in Chicago in regards to the Violation Extradition Act. Within minutes, the subject was identified as wanted by the Sheriff's Office in Purvis, Miss., for homicide since July 17, 2010. This individual had previous arrests in Illinois for manufacturing/delivering less than 15 grams of controlled substances, battery/causing bodily injury, criminal trespass to state land, mob action and manufacturing/delivering a non-narcotic. At 12:54 a.m., a response containing this information was sent to the submitting and wanting agencies.

Host Nation Hire

In 2007, a young man was enrolled into the DOD ABIS as a Host Nation Hire outside the continental U.S. Later that year, he was enrolled again as a linguist. By 2009, when he was biometrically encountered a third time during a badge request, he used a different name, had lost weight and had changed gender from male to female. Despite using a false name and significantly changing his appearance, the individual's identity was revealed by facial recognition and fingerprint matching, proving the value of tracking more than one modality to be able to answer both questions: "Who are you?" and "Have we seen you before?"

---

16 See *infra*, section 3.3.

Delano, Calif.

On Aug. 2, 2010, at 6:30 p.m., IAFIS received an electronic fingerprint submission from the State Prison in Delano for a subject arrested for purchasing/possessing cocaine. Within minutes, the subject was identified as wanted for homicide by the Las Vegas Police Department since April 6, 2010. This individual had previous arrests in California for drug possession, gang membership and robbery. The subject was also arrested in Nevada for domestic battery, grand theft, being a fugitive from justice and trafficking cocaine. At 6:41 p.m., a response containing this information was sent to the submitting and wanting agencies.

Florida Highway Patrol

On Aug. 26, 2011, at 8:09 p.m., a Florida Trooper observed a 1997 Lexus driving without headlights southbound on Interstate 95 near Ormond Beach. When he approached the car, he smelled marijuana. The driver produced a South Carolina driver license, but the alert trooper noticed the driver also had a bankcard bearing another name. The subject's fingerprints were scanned using a mobile ID device.

Within seconds, a hit was returned from the FBI RISC identifying the driver. Gwinnett County, Ga., had issued a warrant for his arrest in connection with a murder and aggravated assault. The warrant had been outstanding for eight years. The trooper also found 9.8 grams of marijuana in the car. The subject was arrested and charged with possession of cannabis, driving with an expired driver's license and being a fugitive from justice.

Of all these elements, establishing the individual's identity (identity proofing) is likely to be the most expensive to develop and maintain. For many small businesses, the cost of developing biometric identity authentication for e-commerce is likely to be unacceptable.

As part of the identity proofing process, it is necessary to collect and process a substantial amount of personal data. Once collected and processed, this data can have great value for marketing and advertising. The availability of personal data for these purposes provides a strong incentive for not sharing with competing businesses. Large retailers can add identity proofing to existing credentialing processes and offer discounts and services for users willing to give up personally identifying information. For small businesses, the cost of collecting customers' data and establishing a database may prove prohibitive. To lower the cost to an individual business, it may be possible to develop new business services that offer identity proofing to retailers on a fee basis, much like the credit card business. However, this would require the establishment of a new business model with associated institutional and cost barriers to entry.

The third factor is the cost of replacing legacy identification processes. Most institutions that might benefit from reduced fraud associated with the use of credit cards by credentials with biometrics would need to replace their current protection framework. For example, financial institutions have been using passwords and bankcards for a long time; they have excellent actuarial knowledge of potential fraud and abuse and have mechanisms in place to protect themselves from expected losses. Introducing a biometrically protected security mechanism will result in uncertainty. The financial institution, or their insurer, is likely to have trouble estimating the new technology's cost and benefit tradeoffs. While adopting a biometric solution may reduce the risk exposure, the institution will need to know the extent of that change, which will be difficult to estimate. That uncertainty will act to inhibit the technology's adoption.

According to *NIST Special Publication 800-63, Electronic Authentication Guideline* and its draft revision, biometrics are currently not suitable for use in remote electronic authentication. Experts working in the intersection of biometrics, identity management and cryptography have identified two critical gaps that must be addressed before biometrics could be considered as an authentication factor for remote transactions over untrusted networks.

First, biometric sensors need to provide reliable evidence of authentic biometric capture — often referred to as liveness detection. Using a biometric "spoof" is analogous to making an unauthorized copy of a token. Likewise, a biometric sensor's physical and digital integrity must be analogous to the various integrity controls with a smart card and its reader.

Second, biometric data must be protected in a renewable and revocable form. Modern cryptography has enabled techniques that allow passwords to be validated without storing the unencrypted password. Secure biometric template technologies (known generally as biometric template protection) would afford similar protections, making it possible to perform biometric matching without requiring knowledge of the underlying biometric data.

To mitigate these constraints a number of measures are needed. One is the availability of sufficient data to compute costs and benefits of biometric services accurately. Since reliable performance data can only be obtained from sources other than the technology provider, government support and certification of appropriate testing procedures and testing organizations is required. In addition, the government will need to support the development and establishment of policies that address personal data ownership and use in a manner that provides incentives and protection for sharing identity data services by multiple users. Finally, a method for protecting biometric data in a renewable and revocable form must be developed.

3.3 Privacy, Civil Rights and Civil Liberties Protection

The benefits of biometric technology present both increased identity protection and risks to privacy, civil rights and civil liberties. The key to addressing these opportunities and risks lies in developing robust and sustainable solutions.

To protect an individual's identity in this fast moving environment, technology and policies that protect the privacy, civil rights and civil liberties of individuals must advance at an equal pace. As President Obama recognized:

> *The United States faces the dual challenge of maintaining an environment that promotes innovation, open interconnectivity, economic prosperity, free trade and freedom while also ensuring public safety, security, civil liberties and privacy.*[17]

The promise of new, groundbreaking applications of biometric technology cannot be realized without corresponding technology and policies to protect privacy, civil rights and civil liberties. For instance, a theft of biometric information could facilitate criminal access to bank accounts and credit cards, allowing the possibility of other criminal activities. Therefore, government and industry are challenged to create smart solutions that allow for the use and sharing of biometrics without creating more risks to individuals such as identity theft.

Government agencies, commercial organizations and academic institutions that collect biometric data assume the responsibility to govern how it is used, retained and shared. Individuals voluntarily providing their biometrics trust that those collecting personal data will live up to their legal obligations to use them in a manner that preserves anonymity when personally identifiable elements are not necessary. It is the biometric collectors' responsibility to carefully determine the minimum biometric data necessary for each situation and to use the least invasive method. If hand geometry is an option, then that method is generally preferable to DNA or fingerprint collection.

To ensure that protections are realized in this rapidly evolving environment, it is critical that researchers devote attention across the full range of biometric applications, including methods to use biometric technology to protect individual privacy, civil rights and civil liberties.

The most pressing challenge facing the biometrics community is to empower organizations and individuals to benefit from the unique advantages biometrics offer and limit the associated risks arising from the inherent uniqueness of biometric

Identity Proofing The process of collecting, storing and maintaining all information and documentation required for verifying and assuring an applicant's identity. Details describing identity proofing for the use by the government is defined in FIPS 201. The process provides the minimal functional and security requirements for achieving a uniform level of assurance for PIV identity credentials. A process similar to one defined in the FIPS 201 may need to be defined for the commercial arena.

---

17 *Cyberspace Policy Review, Assuring a Trusted and Resilient and Communications Infrastructure* (2009): 13. Available: http://www.whitehouse.gov/assets/documents/Cyberspace_Policy_Review_final.pdf.

Orange County, Fla. On Jan. 24, 2009, at 2:09 p.m., IAFIS received an electronic fingerprint submission from the Orange County Sheriff's Office for an individual who had been arrested for a non-moving traffic violation and driving with a suspended license, first offense. The fingerprints were processed on IAFIS and, within two minutes, the individual was identified as wanted by the Sheriff's Office Marietta, Ga., for burglary since April 30, 2007. The individual had previous arrests in Georgia and Florida including two counts of theft by taking; two counts of giving a false name, address or birth date to law enforcement offices; willful obstruction of law enforcement officers; probation violation; possession of less than 1 ounce of marijuana; failure to appear; purchase, possession, manufacture, distribution or sale of marijuana; and battery/family violence. The individual used a false name at the time of arrest. At 2:11 p.m., a response containing this information was sent to the submitting and wanting agencies.

information. The research community can facilitate smart use of biometrics by creating new methods to safeguard and control their use. Technology and policy research in the following areas will help advance the understanding and ability to use biometrics appropriately.

- **Cancellable Biometrics:** While great strides can be witnessed in anonymization and de-identification research, more needs to be done in developing biometric template protection (also known as cancelable or revocable biometrics) so that re-issuance of a new credential functions similar to conventional passwords. Inspired by the challenge to create cancellable identifiers, industry has begun creating "template protection," which permits the issuance of multiple "distorted" unique templates that are associated with an enrolled image. For instance, if a template is compromised through a data breach or cyber intrusion, then the affected template can be cancelled, and a new one can be issued without sacrificing individual rights, matching performance or data integrity.[18]

- **Integration of Biometric Identity Services Into Online Identity Platforms:** As evidenced by the recent announcement of the *National Strategy for Trusted Identities in Cyberspace*,[19] online identity systems are becoming a critical component of commerce and government operations. There are many instances when an individual has a legitimate expectation of anonymity and should not have to self identify. Therefore, biometric applications should enable people to emerge from anonymity to interact with a system for a specific service and then return to anonymity. The biometrics research community can support the public and private sectors through further developing ways to enhance privacy, security, interoperability and ease-of-use.

- **Ask the Right Questions:** The most helpful contribution to this area of research is to improve upon the research questions themselves. The social and legal implications of biometrics are intimately connected to the capability of the science and technology of biometrics. As the biometrics research community's ingenuity leads toward innovations, they must continually question how each new advance will affect privacy, civil rights and civil liberties. For instance, can true biometric anonymity be achieved that prevents a solution from being reengineered to reveal the underlying individual? Can it be ensured that any race, ethnic or medical indices on a biometric are concealed and not used for discriminatory purposes? These and many other provocative questions must be asked by the legal, policy, research and science sectors of the biometrics community to ensure the protection of the public good is sustained as new benefits of biometric technology continue to be discovered.

---

18 Dr. Andrew Teoh Beng Jin and Lim Meng Hui, *Cancelable Biometrics* (2010). Available: http://www.scholarpedia.org/article/Cancelable_biometrics.
19 *National Strategy for Trusted Identities in Cyberspace, Enhancing Online Choice, Efficiency, Security and Privacy* (April 2011). The report defines a set of guiding principles for achieving a successful and ideal identity ecosystem strategy.

3.4 Interagency Cooperation and Partnering

The Subcommittee has been very successful in coordinating biometrics-related activities of interagency importance. This has been realized in great part through the consistent and long-term multi-agency support of standards development and technology evaluations conducted by NIST. NIST has played a significant role in the improvement of biometric standards, algorithms and products through testing programs that provide a forum for the evaluation of technology performance, interoperability and usability. This impact has been realized through Subcommittee support with member agencies providing operational requirements and gaps for consideration in experimental design, large sample populations of secured de-identified biometric samples for testing and supplemental funding to extend the scope of work. Through support of NIST, the Subcommittee has leveraged combined resources and in return obtained performance assessments of technologies used in very large-scale biometric systems such as IDENT, ABIS and IAFIS/NGI; received results from operationally relevant benchmark studies to support decision makers and procurement officials who would otherwise be left to work with unverified claims of performance and capability; and mobilized the biometrics industry and research community through coordinated challenge problems and evaluations to address and fill important gaps in technology performance and interoperability standards. Since 2006, this support enabled NIST to conduct the following evaluations:[20]

- Multiple Biometric Grand Challenge
    - Still Face Track
        - 1:1 Matching
        - Large Scale 1: Many Face Recognition
    - Iris Portal Track
        - Stand-Off Iris Matching
    - Video Track
        - Face Recognition From Video

- Iris Exchange (IREX) Testing
    - IREX I: Compact Iris Records
    - IREX II: Iris Quality Calibration and Evaluation
    - IREX III: Large Scale 1:Many Iris Recognition

- Face and Ocular Challenge Series
    - The Good, Bad and Ugly Still Face Track
    - Video Track
    - Ocular Track

- Proprietary Fingerprint Template (PFT)
    - PFT II: 1:1 Proprietary Fingerprint Matching

- Minutiae Exchange (MINEX)
    - MINEX II: Match-on-Card Technology
    - Ongoing MINEX: Standard Minutiae Template Interoperability

San Diego

On June 3, 2010, at 8:09 a.m., IAFIS received an electronic fingerprint submission from the Sheriff's Office in San Diego for an inquiry. Within minutes, the subject was identified as wanted by the New York Police Department for homicide. This individual had previous arrests in New York for unlawful assembly, possession of marijuana, murder and gang assault. The individual was also arrested in Louisiana for presence in the United States without permission and in California as a fugitive from justice.

---

[20] For a more complete list of NIST biometric projects and technology evaluations, go to http://biometrics.nist.gov and http://www.nist.gov/itl/iad/ig/biometric_evaluations.cfm.

4. User-Centric Biometrics Approach

Texarkana, Texas On Jan. 6, 2011, at 6:54 p.m., IAFIS received an electronic fingerprint submission from the Federal Correctional Institute in Texarkana with an inquiry. Within minutes, the subject was identified as wanted by the Sheriff's Office in Lawrenceville, Ga., for rape (strong-arm) since Nov. 5, 2009. This individual had previous arrests in Florida and Georgia for disorderly conduct, indecent exposure, child molestation and possession of child pornography.

Washington, D.C. On Aug. 9, 2010, at 11:14 a.m., IAFIS received an electronic fingerprint submission from Interpol in Washington for an inquiry. Within minutes, the subject was identified as wanted by the Los Angeles Police Department for homicide since Dec. 16, 1998. This individual had previous arrests in California for taking a vehicle without the owner's consent, possession of a dangerous weapon, tampering with the marks of a firearm and burglary. At 11:40 a.m., a response containing this information was sent to the submitting and wanting agencies.

Efforts to meet government and private sector biometrics needs will be undertaken in the context of the Nation's complex and continuously evolving society. Perhaps for the first time in the post-industrial, technology driven, information age, societies are not just reacting to technologies but shaping them on a global level. Enabled by the convergence of the advanced internet, mobile communications and computing technologies, intuitive, easy to use mobile devices with an expanding suite of sensors (voice, camera, accelerometer, GPS, etc.) are now ubiquitous on a global scale.

Social media is one dimension of a symbiotic relationship emerging between technology and society that shapes how people live, perceive and identify with each other. Even more so than at the time of the 2006 *Challenge*, this new context requires that the human be at the center of the design of human biometric recognition systems. Research challenges in biometrics are best viewed from the socio-technical wave occurring now and that will grow over the period of this document. This wave will shape expectations for technology and determine the acceptance of the extent of its role in individuals' daily lives.

The increasing availability of mobile, universal computing and communication platforms coupled with users' expectation of convenient and secure applications will drive the development and acceptance of biometric systems in the commercial sector over the next 10 years. Rapidly increasing wireless connectivity and bandwidth coupled with cloud computing paradigms will render mobile devices as the preferred means to access services and interact with private and government entities. Users' mobile appliances will serve as a platform for applications operating among multiple systems that can preserve anonymity when desired, provide secure user authentication for trusted transactions and still allow for forensic analysis of transactions under judicial authority when cause is shown.

In this customer-driven market — characterized by large volume, low margin and rapidly changing cutting-edge technological advantages — mobile applications requiring multifactor authentication will become commonplace for the average user. To thrive, existing and future biometrics modalities need to achieve customer acceptance and trust, and these modalities will become part of the mobile multifactor authentication mix by virtue of the value they offer. Despite the fundamentally different application domain of government biometric systems, this technological wave will inexorably raise civil and military users' expectations of government-sector biometric systems, driving system design and implementation.

The needs this document outlines will most effectively be met by biometrics riding this global socio-technical wave rather than following in its wake. To achieve these goals, the system design will place users who provide their biometrics at the center. User-centric, system-level design strives foremost to understand users, their interaction with the system via usability testing and its impact on performance.[21] Trust in biometrics technology will be earned by virtue of system performance consistently meeting user expectations.

---

21 See e.g., http://zing.ncsl.nist.gov/biousa. NIST biometric usability studies have tested the effect of scanner height and angle on fingerprint capture, measured the effect of repeated users (habituation) and evaluated various methods for conveying biometric collection instructions to a diverse multilingual set of users.

5. Science and Technology

5.1 Research Challenges

Effective use of biometrics over the coming decade, whether in commercial or government domains, requires that research challenges be in key areas:

- **Fundamental Underpinnings:** The fundamental understanding of biometrics must be solidified to obtain biometric measures on a more scientific basis, address various operational and environmental conditions, and model and scale human-machine systems with predictable performance.

- **Biometric Capture:** The ability to capture biometric data quickly and accurately must continue to improve across a range of challenging environments, from unhabituated users in a commercial application to non-cooperative and uncooperative users in the battlefield environment.

- **Extraction and Representation:** Extract and represent biometric data to maintain individuality and efficiently achieve automated retrieval, recognition and interoperability across devices and systems.

- **Trusted Systems:** Design for acceptance through understanding the socio-technical basis of user expectations, securing data and systems effectively against vulnerabilities, establishing a chain of trust for biometric data and enabling its revocation when necessary.

- **Privacy:** Advance technology and policy to enable robust and sustainable solutions that empower organizations and individuals to benefit from the unique advantages biometrics offer while limiting the associated risks to privacy and civil liberties.

- **Standards and Testing:** Address challenges in order to provide the proper focus to continue the advances made during the previous challenge period.

5.2 Research Focus Areas

The recommended research foci presented are based on the analysis provided in two key National Research Council (NRC) reports, a National Science Foundation (NSF) workshop, two workshops held by the Biometrics and Identity Management Subcommittee in support of this *Challenge* update and a review of published materials. The two key reports are the *Biometric Recognition Challenges and Opportunities*, published by the NRC in 2010, and *Strengthening Forensic Science in the United States*, published by the Committee on Identifying the Needs of the Forensic Sciences Community of the NRC in 2009.

The first report addresses the need for greater knowledge of biometric characteristics and calls for improved systems approaches to the development of recognition systems. It also adds to the understanding of privacy and legal issues related to biometrics. The second report is a response to the misidentification of the Mayfield[22] fingerprint. This misidentification, against a background of more frequent

---

Intelligence Advanced Research Projects Activity (IARPA) Since Sept. 11, the need for reliable biometric recognition performance has expanded beyond access control and verification applications that operate within tightly controlled conditions.
The IARPA Biometrics Exploitation Science and Technology (BEST) Program is conducting high-risk, high-payoff research to advance biometrics technology to meet these challenges.

IARPA's BEST Program goals are to: 1) significantly advance the ability to achieve high-confidence match performance, even when the biometric features are derived from non-ideal data; and 2) significantly relax the constraints currently required to acquire high-fidelity biometric signatures.

Columbia, S.C.

On March 17, 2010, at 10:03 a.m., IAFIS received an electronic fingerprint submission from the Department of Corrections-Emergency Action Center in Columbia for assault/battery. The fingerprints were processed on IAFIS. Within minutes, the subject was identified as wanted for rape by the Sheriff's Office in Hagerstown, Md., since Oct. 5, 2006. The individual had previous arrests in Maryland for theft, assault, robbery, destruction of property, and possession of cocaine with intent to distribute. The subject also has arrests in South Carolina for criminal sexual misconduct and kidnapping. At 10:14 a.m., a response containing this information was sent to the submitting and wanting agencies.

---

22 DOJ Office of the Inspector General, *A Review of the FBI's Handling of the Brandon Mayfield Case: Executive Summary* (January 2006). An excerpt: "In May 2004, the FBI arrested Brandon Mayfield, an Oregon attorney, as a material witness in an investigation of the terrorist attacks on commuter trains in Madrid, Spain, that took place in March 2004. Mayfield had been identified by the FBI laboratory as the source of a fingerprint found on a bag of detonators in Madrid that was connected to the attacks.

Houston

On Oct. 16, 2009, at 10:46 a.m., IAFIS received an electronic fingerprint submission from the Sheriff's Office in Houston for aggravated robbery and assault against a public servant. The fingerprints were processed on IAFIS; within minutes, the individual was identified as wanted by the Harvey Sheriff's Office in New Orleans for murder since March 25, 2008. The individual had previous arrests in Louisiana for escape, aggravated battery, armed robbery, kidnapping, possession
of stolen property and parole violation. At 10:50 a.m., a response containing this information was sent to the submitting and wanting agencies.

---

challenges of biometric identifications, resulted in a comprehensive analysis of the state of forensic identification. The report calls for establishing a scientific basis for forensic identification that satisfies the "*Daubert* Criteria."[23] The FBI, the NSTC Subcommittee on Forensic Science and others have initiated studies to address the issues these reports identify.[24] This effort must continue and be expanded to fully address the reports' major concerns.

The November 2010 NSF Workshop on Fundamental Research Challenges for Trustworthy Biometrics brought together more than 50 academic, government and industry experts in biometric systems and cyber-security with the charge of identifying the fundamental research challenges for trustworthy biometric systems. The workshop acknowledged the current government-funded research focus on biometric capture (sensors and systems), application-oriented systems (e.g., border security, wartime detainment) and advanced signal and image processing techniques may add robustness to current systems. The workshop highlighted needs consistent with the NRC report *Biometric Recognition Challenges and Opportunities* and the NSTIC report *National Strategy for Trusted Identities in Cyberspace*. Research must be undertaken that addresses privacy, acceptability, usability and security of stored and transmitted biometric information. Consistent with these findings, scientific fundamentals of identity science, as well as research at the intersection of identity management and cyberspace, must be advanced.

There were two workshop meetings held to support the *Challenge* update. The first was the March 28, 2011, International Biometrics and Identification Association meeting. The workshop's focus was to gather industry input concerning the *Biometrics Challenge Update*. The meeting provided insight into the commercial biometrics marketplace. Key issues identified at the workshop were a need to address public perceptions of privacy, improved security, spoofing detection and issues related to improved data acquisition and extraction.

More than 100 biometric experts attended the second workshop that convened in May; the workshop featured highly interactive discussions on the future of biometrics. The workshop supported the above findings and called for recognizing the need to address the complex nature of privacy, provide improved security based on multi-factor authentication, the development of an identity-proofing framework for use by the commercial marketplace and greater standardization across all biometric applications. Recommended research areas that developed from these workshops are:

Fundamental Underpinnings

- Establish the scientific foundation of intrinsic biological distinctiveness of biometric modalities as influenced by demographic variations;

- Establish the scientific foundation that determines the extrinsic distinctiveness of biometric modalities under a range of different collection modes and environments including human interface, sample image quality, image size and image resolution;

Approximately two weeks after Mayfield was arrested, the Spanish National Police informed the FBI that it had identified an Algerian national as the source of the fingerprint on the bag. After the FBI laboratory examined the fingerprints of the Algerian, it withdrew its identification of Mayfield, and he was released from custody."

23 *Daubert v. Merrell Dow Pharmaceuticals* (92–102), 509 U.S. 579 (1993).

24 An example of the progress in this area is: Bradford T. Ulery, R. Austin Hicklin, JoAnn Buscaglia and Maria Antonia Roberts, *Proceedings of the National Academy of Sciences*, "Accuracy and Reliability of Forensic Latent Fingerprint Decisions" (May 2011).

- Establish the scientific foundation that determines the intrinsic stability of biometric modalities as well as understand the variation in established and emerging biometric modalities (e.g., as a function of time, environment). Specifically, establish the stability of iris and face images as a function of age, environment and other factors; and

- Establish the scientific foundation to determine the measures of comparison of likelihood for all biometric matches taking into account sample image quality, image size and image resolution. Research is necessary to develop:

    - Quantifiable measures of likelihood in the conclusions of forensic analyses; and

    - Measures the likelihood for all automated fingerprint, face and iris recognition systems.

Biometric Capture

- Advance integrated or hybrid-integrated sensor technologies enabling adaptive, near-simultaneous acquisition of co-registered multispectral and/or multimodal data;

- Advance integrated, highly compact and robust imaging systems that are broadly adaptive across multiple spectral bands;

- Develop more robust capture systems for acquiring biometrics from non-cooperative subjects from a distance and while engaged in other activities;

- Develop better understanding of usability factors as they relate to biometric capture devices taking into account the device's design, environmental factors and habituation of user; and

- Research of transformational nanobiotechnology approaches for real-time molecular biometrics based on DNA and new modalities, including sensitive yet selective molecular recognition without molecular amplification.

Extraction and Representation

- Perform fundamental research resulting in the identification of approaches for the robust segmentation and exploitation of human biometric information from the information cacophony characteristic of human environments and activity;

- Develop invariant representations of individuals (e.g., drawing on multiple sensor and/or biometric modalities) that maintain adequate uniqueness/individuality under system scaling and that are robust to change due to intrinsic and extrinsic factors;

- Develop a forensic basis for making forensic comparison decisions between fingerprints captured using emerging collection technology (e.g., contactless) and legacy data and latent fingerprints captured using traditional methods;

- Undertake fundamental research into performance limits/system scalability with matching and indexing optimized for new invariant representations;

Columbus, Ga.

On March 30, 2009, at 7:43 a.m., IAFIS received an electronic fingerprint submission from the Columbus Police Department for an individual arrested for willful obstruction of a law enforcement officer and a family violation. The fingerprints were processed on IAFIS, and, within 56 seconds, the individual was identified as wanted by the FBI in San Francisco for murder since June 9, 2006. The individual had a criminal history in California, Texas and Georgia. The history included previous arrests for false identification to peace officers, two counts of unlawful sex with a minor older than three years, oral copulation with a minor older than 10 years, receiving stolen property, revoked probation, criminal intent to terrorize, two counts of driving on a suspended license, failure to appear, domestic violence, battery on a spouse, child stealing, exhibiting a firearm, carrying a concealed weapon in a vehicle, felony possession of a firearm, purchase and sale of a narcotic substance, and parole violation. The individual used a false name at the time of arrest. At 8:39 a.m., a response containing this information was sent to the submitting and wanting agencies.

Union, N.J.

On Nov. 28, 2008, at 7:15 p.m., IAFIS received an electronic fingerprint submission from the Clark Township Police Department in Union for an individual arrested for theft. The fingerprints were processed on IAFIS. Within 27 seconds, the individual was identified as wanted by the Dallas Sheriff's Office for sexual assault since Sept. 22, 2003. The individual had previous arrests in Texas and New Jersey that included driving while intoxicated and theft. The individual used a false name at the time of arrest. At 7:15 p.m., a response containing this information was sent to the submitting and wanting agencies.

- Maintain and expand test databases for fingerprints, palmprints, face, iris and other biometrics for use in research and technology testing, within the constraints imposed by institutional review boards and individual privacy protection requirements; and

- Develop performance models for the major modalities that effectively estimate performance using smaller test databases that represent different sources, populations and environments.

Trusted Systems

- Establish a framework through which the societal, legal and technological dynamics impacting biometrics usage may be understood and system design principles established;

- Develop formal system-level design methods for biometric systems;

- Conduct system security, integrity, reliability and acceptance research with outcomes advancing design principles for realization of trusted biometric systems;

- Establish via system-level research the role of trusted biometrics in multifactor authentication systems for securing identity in cyber systems and how these systems may be understood, specified and assessed;

- Develop metrics for liveness detection and evaluation methodologies at the capture subsystem; and

- Develop testing metrics and methods for biometrics template protection.

Privacy

- Support fundamental research aimed at explicating "privacy" as various segments of the population understand it;

- Integrate privacy principles into the design of biometric systems and their different stages, including enrollment, storage, capture and transport;

- Identify privacy issues as they relate to potential use of identity data by service providers of identity proofing services; and

- Identify privacy issues as they relate to commercial business use of biometric data and associated identity data.

Standards and Testing

- Develop best practices and standards to support large-scale framework for e-government, personal information and business transactions;

- Continue support for multi-factor verification and authentication that includes biometrics;

- Develop standard data interchange formats for information on liveness detection, both between modules and between systems;

- Develop standards for revocable biometrics (biometric template protection);

- Provide support for ongoing programs to develop fraud detection standards and develop evaluation methods for fraud detection;

- Fully develop and adopt a government-wide person-centric identity management data model. The National Information Exchange Model is a step in this direction;

- Continue development of biometrics system performance testing standards;

- Continue development and standardization of image quality metrics for face and iris;

- Define and standardize "plug-and-play" interfaces and software practices;

- Provide continued Standards Developing Organization support including developing reference implementations, conformance test suites and testing of standards prior to publication. The Committee to Define an Extended Fingerprint Feature Set is a prime example of such testing prior to publication;

- Provide institutionalized support to government testing entities to develop certification programs;

- Conduct technology testing for operational effectiveness, suitability and interoperability;

- Develop a framework for collecting test data on an ongoing basis and developing provisions for making the data available to independent testing entities; and

- Develop a framework for a coordinated and fully cross-referenced list of approved and or certified biometric products for use across all Federal agencies.

5.3 Education and Training

As biometric system development and deployment rapidly grew to meet the needs the 2006 *Challenge* stated, experts quickly recognized that individuals with adequate educational training or experience needed to staff these collection and analysis efforts were in short supply in the private and public sectors. Research and user-centric biometric system designs increasingly require a highly interdisciplinary skill set. Educating and training an effective workforce of biometric professionals at all levels remains a critical requirement in 2011.

Maintaining globally competitive education and training in biometrics is central to achieving the challenges this document lays out. The *Challenge* recommends education and training, including:

- Building on the IEEE Certified Biometrics Professional efforts to date, the biometrics community should consider regularly revisiting and updating the biometrics body of knowledge. Because the technology and the societal context it is used in are constantly evolving, this is an ongoing challenge. However, standardization is essential to regularly consolidating the scientific advances and determining the basic knowledge required by professionals in this field.

Latent Fingerprint Identifications

On Jan. 27, 2009, BIMA examiners confirmed a possible match of latent fingerprint identifications from two IED-related incidents that occurred in different locations more than 15 months apart. The new generation ABIS makes building these links possible through latent-to-latent matching, and it provides key intelligence that helps makes sense of the whereabouts and patterns of individuals that elude U.S. forces. An unknown target's anonymity lessens a little bit each time he leaves a fingerprint behind.

East Baton Rouge Parish, La.

A subject was arrested for three counts of burglary and one count of illegal possession of stolen property. His fingerprints check revealed that he had used at least three aliases, was an aggravated felon and had been previously removed. He had been sentenced to five years' incarceration for manufacturing methamphetamine and felony first-degree forgery. Upon disposition of his current charges, he will be held in ICE custody pending removal.

Yuma, Ariz.

In August 2010, the Border Patrol in Yuma apprehended a man who had entered the United States illegally. The Biometrics Security Consortium ran his fingerprints against IDENT and determined, through interoperability with the FBI IAFIS that he had two outstanding warrants, including one for homicide and was considered armed and dangerous. He was taken into custody and faces charges in the 2004 stabbing death of his girlfriend in Oregon.

- Since biometrics encompasses multiple fields, an integrated, comprehensive academic curriculum is necessary to achieve required levels of proficiency. Pertinent topics for integrative treatment include identity management, sensor design, pattern recognition, computer vision, signal processing, applied biology, applied mathematics, industrial statistics, forensics, privacy, security, international law, database design, management information systems and economics. Such curricular innovations would help to create further biometric programs for undergraduate, graduate and post-graduate students. These curricula would provide a generation of the biometrics workforce with a holistic understanding of the technology.

- The robust research required to meet the biometrics community's needs requires a vibrant and diverse academic research community cooperatively engaged with industry and government. Training students at all levels through this research will provide the next generation of practitioners, innovators and teachers. Robust, consistent research funding in the challenge areas is essential to maintaining this community and its progress. Biometrics is very much an application-specific field. Partnerships between academic programs, industry and government are essential to enable new graduates to have real experiences that prepare them to function effectively in their new jobs.

- Some individuals and organizations view biometrics as an invasive technology that systematically violates the individual's privacy. This fear is often based on a false understanding of its capabilities and applications. A concerted dialogue is needed to engage and properly educate society about the technology and the privacy protection capabilities of systems that use biometrics.

6. Standards, Commercial Testing and Certification

The early advance of biometric technology, to a large extent, has been funded and driven by government to meet critical mission needs of individual departments. The pioneers and early adopters were law enforcement agencies with fingerprint collections that had grown too large to be efficiently and cost effectively maintained manually. These agencies wanted to stimulate the development of biometric technology and then have it adopted, commercialized and eventually maintained and improved by industry. In this they succeeded; perhaps too well. Multiple, mutually incompatible, products were developed by competing firms and purchased by state and local governments. The cost of the systems was high and usually served to lock the agencies into reliance upon a particular vendor for very long periods. In time, it was recognized within law enforcement that islands of local and state biometric automation, unable to interoperate with neighboring jurisdictions were not in anyone's interest. Coinciding with development of a national identification and criminal history system by the FBI, standards were developed and adopted and included by law enforcement agencies in subsequent system acquisitions to address the problems found. But, the lack of a priori standards has had an adverse impact, particularly upon latent fingerprint matching, which has continued up until the present.

As biometric systems were developed to support visa issuance, border control and military operations, the need for standards was recognized but too often occurred without recognizing a need for all of the national systems to interoperate. One of the positive outcomes of the NSTC Subcommittee on Biometrics and Identity Management has been an increased emphasis upon standards and processes to insure that they are followed.

The use of standards contributes to product maturity by providing many benefits to industry and biometric end users, including:

- **Reduced Product Cost:** By manufacturing to a standard, economy of scale can be realized because of cheaper supplier prices from the vendor, fewer variations of model design and more efficient use of the means of production. Other factors contributing to lower cost are reusing testing procedures and certifications.

- **Decreased Implementation Time for the System Developer:** Standards promote interconnectivity among products and components, product reuse, shorter delivery times and a reduction of the time required for design, integration and testing.

- **Better Selection for the Product User:** Standards allow the program developer to select the best product and not be locked in by proprietary design. Likewise, vendors are encouraged to compete based on innovation, quality and reliability.

- **Greater Resource Sharing for Identification:** Standards for data exchange between biometrics systems support more effective resource utilization.

- **Less Expensive, More Rapid Technology Improvement:** Standards provide a clear definition of needs to industry, facilitating product improvement programs.

Gwinnett County, Ga.

A subject was arrested for violating probation, which he received for a felony residential burglary conviction. Upon booking, he claimed to be a U.S. citizen and was not brought to the attention of ICE agents. However, his fingerprints check revealed a criminal history spanning 10 years and two states. His prior convictions included felony battery on law enforcement, felony cocaine possession, felony habitually driving without a license, battery and family violence. He had used at least 15 aliases during 25 previous encounters with law enforcement. ICE will remove him from the United States upon completion of his criminal sentence.

> **Boston**
> A subject was arrested and convicted for felony assault and battery. While in law enforcement custody, her fingerprints were checked against DOJ and DHS records. ICE was immediately notified that she had illegally entered the United States in 2005 and had an order for removal after failing to appear before an immigration judge. ICE removed her from the United States.

- **Consistent Performance Expectations:** Standards help create consistent user expectations of performance across different groups and organizations.

- **Easier System Integration:** Standards support integration of disparate systems produced by different vendors.

6.1 Accomplishments

Since 2006, the biometric standards landscape has changed dramatically. Older standards have been expanded and modified; new standards have been created in several areas. The main accomplishments over the last five years are:

- NSTC published the *U.S. Government Recommended Biometric Standards*. This document, which is updated periodically, provides a list of the standards that Federal biometric systems are expected to comply with.

- The ANSI/NIST-ITL[25] base standard has been expanded to include:

    - Image quality requirements and segmentation data to support the processing of "flat" or plain fingerprint images;

    - Definitions for a new block of minutiae fields to harmonize with the International Committee of Information Technology Standards (INCITS) M1 minutiae standard;

    - Adoption of provisions for use of variable resolution records;

    - Best practice application levels for the capture of facial images;

    - A new record type for the exhange of iris information;

    - Publication and inclusion of the fingerprint Extended Feature Set standard defining a vendor-neutral, feature definition format;

    - A new record type to contain biometric information not described in this standard but that conforms to other registered biometric data format standards; and

    - An XML alternative representation for this standard.

- Agreements have moved fingerprint interoperability forward in the European Union, thereby contributing to greater globalization of the biometric interchange standards.[26]

- Internationally accepted and implemented travel document standards have been developed, including *ICAO 9303* and periodic supplements.

- The FIPS 201 PIV program has led to the standardization of biometric processing for electronic credentials used for logical and physical access control systems, such as:

---

[25] 2011 will see another major update to the ANSI/NIST standard for biometric data transmission with the creation of additional records to provide sharing of DNA data, original source images, information assurance and both traditional and XML encodings as informative annexes.

[26] Prüm Convention, *Council Document No. 10900/05 and Implementing Agreement, Council Document No. 5473/07*. A convention between the Kingdom of Belgium, the Federal Republic of Germany, the Kingdom of Spain, the French Republic, the Grand Duchy of Luxembourg, the Kingdom of the Netherlands and the Republic of Austria on increasing cross-border cooperation, particularly in combating terrorism, cross-border crime and illegal migration. The Prüm Convention occurred May 27, 2005, and chapter two of the convention addresses DNA profiles, fingerprinting and other data.

- Fingerprint template standardization and certification;
- Iris; and
- Facial image.

• Using the National Information Exchange Model in ANSI/NIST-ITL standard facilitated information exchange between diverse users.

• NIST has developed the *Mobile ID Device Best Practice Recommendation (Special Publication 500-280)*, which codifies best practices for handheld biometric devices for fingerprint, face and iris modalities.

• Translators have been developed and implemented for exchanging data between systems using different data exchange specifications between virtually all of the large national and international biometric identification systems (CJIS IAFIS and US-VISIT IDENT, CJIS IAFIS and INTERPOL systems, and CJIS IAFIS and DOD ABIS).

• Numerous additional standards have been created for other modalities and applications.

• The second-generation revision process for ISO standards[27] are coming to a close.

Significant progress has also been achieved in the testing and certification of key biometric products. NSTC publishes and maintains a catalog of Federal Biometric Testing Programs that provide references to these products, including:

• A FIPS 201 testing and certification program with a public website listing several hundred biometric products that GSA and other supporting organizations, such as NIST, have tested and certified.

• FBI's EBTS Appendix F fingerprint image quality certification program for fingerprint scanners, card readers and mobile devices also lists hundreds of qualified products on its website.

• The WSQ fingerprint compression algorithm certification program is now established.

• An extensive DOD program of biometric product testing and certification has been established.

• Conformance testing methodologies have been developed for a variety of standards.

In addition to Federal testing programs, the NIST National Voluntary Laboratory Accreditation Program has begun certifying independent laboratories to provide third party testing of biometric products within well-defined scopes of testing.

Some certification programs listed above are limited to specific capabilities. For example, testing to Appendix F specifications ensures that the image quality is compliant, but it does not address any of the scanner's other functionalities — such as data entry, usability or physical requirements. More importantly, no program currently provides a cross-reference of standard compliance that can link the various testing programs for the various Federal agencies. This means, that all too often,

Lake County, Ill.

On May 22, 2010, ICE agents encountered a male subject who was detained at the Lake County jail after being arrested for resisting a peace officer. IDENT/IAFIS Interoperability assisted in identifying him as a native of Mexico who has used at least 11 aliases during previous encounters with law enforcement. He was first removed from the United States in 1989, and he has since been removed three times. In 2003, he was convicted of illegal reentry after deportation and sentenced to 90 months in prison. He was also sentenced to four years' incarceration for an aggravated felony burglary conviction, and has been convicted of multiple vehicle-related misdemeanors. ICE reinstated his prior deportation order and removed him from the United States on July 13, 2010.

---

27   ISO/IEC JTC 1/SC 37. Available from: http://www.iso.org/iso/jtc1_sc37_home.

London, Ohio

On March 12, 2010, at 3:33 p.m., IAFIS received an electronic fingerprint submission from the Ohio Bureau of Criminal Identification in London for Child-Adult Care/School Employment. The Non-Federal Applicant User Fee fingerprints were processed on IAFIS. Within one minute, the individual was identified. This individual had previous arrests in Ohio for nighttime breaking and entering and domestic violence.

---

the manufacturer must design and test products to provide similar functionalities to satisfy differing, but often nearly identical, requirements for various agencies.

6.2 Trends and Challenges

Acceptance and use of biometrics by the U.S. and international law enforcement, border and immigration management agencies have provided a major impetus for development in the biometrics industry. Currently, there is a growing interest in using biometrics in the private sector, particularly in the health care industry and financial institutions. Interest in other public sectors has also increased. Public agencies dealing with benefit management, licensing and e-government also show a growing interest. Concurrently, there is a growing interest in applying biometrics to these newer applications and an explosive growth in inexpensive, fast communications devices with biometric capabilities.

The proliferation of these biometrics-capable, multi-functional devices will further contribute to a growing demand for biometric services. Combining demand for biometric services and the ability to deliver them with the aid of inexpensive biometrically enabled devices poses a significant challenge. The increase in demand will be further driven by the need for greater security and privacy protection. Standards addressing these types of applications will need to be strengthened and, in some cases, developed.

There has been unprecedented growth of national identification systems. DOD's wide use in-theater during military operations and the growing use of biometrics in border and immigration management systems have sometimes caused interoperability to be a casualty of the technology's rapid progress. Agencies are focused on using the ANSI/NIST-ITL standard as the basis for their transactions. With this standard, they are developing application profiles to extend the basic interoperability needs for interagency transfer to meet their own internal needs. The core set of information needed for interoperability has now been agreed on and is being implemented. With the pending adoption of the ANSI/NIST-ITL version 1-2011, the agencies' interoperability capabilities will expand to include more forensic analysis of biometric data. It will also formalize the processes and methods to exchange information about DNA. There has been a high level of interaction among affected government agencies (local, state, national and international — such as foreign governments, INTERPOL and the European Union) while updating the ANSI/ NIST-ITL standard.

Standardization of biometric software is increasing. However, it has not achieved the same level of interoperability as hardware. Initiatives to standardize software interfaces have been attempted, with limited success. An example of an ongoing project is the Biometric Identity Assurance Services (BIAS) standard being developed under the Organization for the Advancement of Structured Information Standards (OASIS).

BIAS defines a framework for deploying and invoking biometrics-based identity assurance capabilities that can be readily accessed using services-based frameworks. The BIAS standard, a Simple Object Access Protocol (SOAP) Profile aligning to INCITS 442:2010, offers a Web Services Description Language document describing available operations. This standard has been in development for more than five years and is nearing completion. The Common Biometric Exchange Formats Framework (CBEFF) standard provides the ability to identify different biometric data structures (public or proprietary) supporting multiple biometric devices and applications. CBEFF also enables the exchange of biometric information efficiently between system components; it is gaining increasing acceptance in large-scale applications of biometrics in international applications such as the UIDAI project. Acceptance in the U.S., however, has been limited.

Implementation of SOAP using Web services has resulted in modularized software design. However, this has been achieved mostly on an ad hoc basis. While Web services provide the potential for widely interoperable biometric device interfaces, there are few standards for "plug-and-play" biometric interfaces and software. More research and coordination with industry is required.

Matching algorithm-independent image quality metrics are recognized as an essential element for achieving high comparison accuracy. Algorithms for computing image quality metrics and their inclusion in standards is still evolving, and progress is not uniform across modalities. The NIST NFIQ metric for fingerprint image quality is used widely, but it is not universal and is currently undergoing revision. Alternative commercial fingerprint image quality tools are available and used by many applications. Face and iris image quality standards for applicability across a range of algorithms are relatively immature.[28]

Security concerns regarding the use of biometrics have increased significantly over the last five years. There is a growing awareness of more frequent occurrences of biometric fraud. Algorithms and software have been developed to flag potential fraud, but these countermeasures are not yet a part of mainstream applications. Work has started on developing standards specifying methods and performance levels for spoofing and avoidance, however, completion will take some time. Compromising a biometric characteristic can have serious consequences to its owner, and methods to protect the biometric characteristic and/or revoke the features derived from that characteristic are desired. Of course, once the biometric characteristic itself has been compromised, there can be no revocation. There has been a growing awareness of the need to address these risks, and proposals for mandating multi-factor authentication have been made to mitigate the risk. This does not directly address the need for developing effective countermeasures to combat spoofing.

The NRC report, *Biometric Recognition Challenges and Opportunities*, highlighted the concern that a lack of operational test data jeopardizes users' confidence in the accuracy of biometrics results. Additionally, the report asserts that the accuracy requirements vary with the mission set employing the particular modalities. Additional operational testing is required to determine system accuracy, generate operational test databases and communicate this information to users to build their assurance of biometrics results.

U.S. government support of standards, testing and certification has contributed to the development of the biometrics industry. Given the large role that government plays in establishing and verifying people's identities and in determining access to sensitive or private data, it is necessary that the government continue to play an active role in developing standards, best practices and standards conformity testing programs. These activities need to be coordinated among the major players to ensure maximum interoperability and product interchangeability to the numerous public and private applications that rely on biometrics.

Heflin, Ala.

On March 3, 2011, at 6:52 p.m., IAFIS received an electronic fingerprint submission from the Sheriff's Office in Heflin for second-degree forgery. Within minutes, the subject was identified as wanted for homicide by the Fulton County Police Department in Atlanta since Dec. 17, 2009. This individual had previous arrests in South Carolina and Georgia for trafficking more than 10 grams but less than 28 grams of ice/crank/ crack, possession of a weapon during a violent crime, cocaine possession and a weapon offense. At 6:55 p.m., a response containing this information was sent to the submitting and wanting agencies.

---

28 For example ISO/IEC 29794:5-2010, *Sample Quality for Face Image Data*, has just been published, however, *Iris Image Quality Standards* have not yet been published.

7. Conclusion

The 2006 *Challenge* outlined the RDT&E priorities needed to meet the Nation's most pressing national security and public safety challenges. Federal agencies, partnering with private industry and academia, followed the RDT&E path laid out in the *Challenge*, which enabled significant advances in operational capabilities. There are still capability gaps that must be addressed. The research needs identified here are the priorities for the next several years. Agency attention and funding are going to focus on these priorities. The next few years will see reductions in agency budgets in many areas. Recognizing this budgetary reality, Federal agencies with major biometric activities are coordinating their efforts and are often including partner agencies in making acquisition decisions, and they will make every effort to prevent duplication of effort. As before, partnership between the U.S. government, the private sector and academia is absolutely necessary for the challenges to be met.

Binghamton, N.Y. On Jan. 25, 2011, at 1:53 p.m., IAFIS received an electronic fingerprint submission from the Sheriff's Office in Binghamton with an inquiry print. Within minutes, the subject was identified as wanted by the police department in Dover, Del., for rape (strong-arm) since Jan. 24, 2011. This individual had previous arrests in Delaware and New York for carrying a concealed weapon, first-degree rape, second-degree rape, second-degree kidnapping, third-degree unlawful sexual contact, offensive touching, first-degree arson, second-degree conspiracy, intentional damage to property, disobeying a court order, aggravated harassment and criminal contempt. At 1:57 p.m., a response containing this information was sent to the submitting and wanting agencies.

NSTC Subcommittee on Biometrics and Identity Management

Co-chair: Mr. Duane Blackburn (OSTP)
Co-chair: Mr. John Boyd (DOD)
Co-chair: Mr. Chris Miles (DHS S&T)

Department Leads

Mr. Jon Atkins (DOS)
Mr. William Baron (DOT)
Mr. Duane Blackburn (EOP)
Mr. John Boyd (DOD)
Ms. Zaida Candelario (Treasury)
Mr. Michael Garris (DOC)
Dr. Larry Hornak (NSF)
Ms. Usha Karne (SSA)
Dr. Michael King (IC)
Mr. James Loudermilk (DOJ)
Mr. Chris Miles (DHS)
Mr. David Temoshok (GSA)

Research, Development, Testing and Evaluation Working Group

Mr. Jon Atkins (DOS)
Mr. William Baron (DOT)
Mr. Sankar Basu (NSF)
Mr. Duane Blackburn (EOP)
Mr. John Boyd (DOD)
Mr. James Buckley (DHS)
Ms. Zaida Candelario (Treasury)
Mr. Semahat Demir (NSF)
Mr. Jeffrey Dunn (IC)
Ms. Valerie Evanoff (FBI)
Ms. Kelly Fadis
Mr. Michael Garris (NIST) – Chair
Mr. Ed German (IC)
Mr. Paul Good (DHS)
Mr. Mark Greene (NIJ)
Mr. Patrick Grother (NIST)
Ms. Heather Haller (FBI)
Mr. Marty Herman (NIST)
Mr. Terry Hess (TSWG)
Ms. Karyn Higa-Smith (CID)
Ms. Usha Karne (SSA)
Mr. Michael King (ODNI)
Mr. Jon Lazar (DOD)
Mr. Dave Lohman (DOD)
Mr. James Loudermilk (FBI)
Mr. Edward McCallum (TSWG)
Mr. Chris Miles (DHS)
Ms. Karen Pate (IC)
Mr. Jonathon Phillips (NIST)
Mr. Peter Sand (DHS)
Mr. Scott Swann (ODNI)
Mr. David Temoshok (GSA)
Ms. Mary Theofanos (NIST)
Mr. Antonio Trinidade (DHS)
Mr. Brad Wing (NIST)
Ms. Patricia Wolfhope (DHS)

Special Acknowledgements

The RDT&E Working Group acknowledges and thanks the following contributors for their efforts in developing this 2011 revision and update of the *National Biometrics Challenge*:

- Dr. Larry Hornak (NSF) and Mr. James Loudermilk (FBI) for co-chairing the effort.

- Ms. Valerie Evanoff, Dr. Larry Hornak, Mr. George Kiebuzinski and Mr. James Loudermilk for providing primary author services.

- BRTRC and FBI CJIS Multimedia Productions Group for editorial services and graphics design.

- The International Biometric Industry Association for hosting an Industry Workshop and other assistance in understanding business needs and issues.

- Ms. Jenna Osteen and Ms. Merideth Cohrs, Booz Allen Hamilton, for organizing and conducting a Government and Industry Workshop.

Tucson, Ariz.

On Nov. 17, 2008 at 3:27 p.m., IAFIS received an electronic fingerprint submission from the Tucson Sheriff's Office for an individual's prints being submitted as a return arrest. The fingerprints were processed on IAFIS. Within one minute, the individual was identified as wanted by the Sheriff's Office, Coeur d'Alene, Idaho, for homicide since Oct. 10, 2008. The individual had previous arrests in Washington, Idaho and Arizona that included possession of a controlled substance, possession with the intent to sell, manufacture of a controlled substance, possession of paraphernalia and possession of a controlled substance with intent to deliver. The individual used a false name at the time of arrest. At 3:28 p.m., a response containing this information was sent to the submitting and wanting agencies.

Appendix – Resourcing the Challenge

Quito, Ecuador

In February 2010, intelligence analysts with the Special Operations Command South deployed to the U.S. Embassy in Quito, Ecuador, to support the ICE Attaché Office during investigations of Ecuador-based Middle Eastern and African smuggling networks. During a two-week deployment, 19 individuals were biometrically enrolled at Quito's international airport. The following June, when two individuals were detained by U.S. Customs Border Patrol officers in the desert near Tucson, Ariz., both were fingerprinted. One provided a "hit" to one of the February Quito enrollments. His travel companion claimed loose ties to Harkat-ul-Jihad al-Islami Bangladesh, a terrorist organization, and both were denied asylum to the U.S.

Thomas Edison's first invention and patent was for an electronic vote recorder. That invention found few takers, and his first venture failed. He would later say of that experience: "Anything that won't sell, I don't want to invent. Its sale is proof of utility, and utility is success."[29] Part of the effort to revise the 2006 *Challenge* entailed meeting with representatives from industry, academia and government and assessing progress made, work remaining to be done and new needs that had arisen. Some feedback from those discussions was that the *Challenge* had been important in directing subsequent research and development efforts.

This 2011 update to the *Challenge* by the Biometrics and Identity Management Subcommittee, while identifying many areas where additional research is needed, does not identify any funding sources. The Subcommittee will not consider any research proposals. Recognizing the uncertainty and constrained budget environment for government, academia and industry over the next few years, the *Challenge* does try to identify and clearly communicate those needs. The *Challenge* does not stand alone in communicating research and development needs.

Many government programs, broad agency announcements and other solicitations exist that further detail the requirements and individual agency mission needs framed within this document. These initiatives, collectively, provide the opportunity for industry and academia to advance discovery and further develop their innovative approaches for satisfying the government's needs. As Edison appreciated, these programs, broad agency announcements and solicitations represent the vehicles for the government to purchase products of demonstrated utility derived from academic and industry research and development investments in biometric and identity management technologies.

Federal Business Opportunities (FedBizOpps)[30] is the single government point-of-entry for Federal government procurement opportunities over $25,000. It provides the public with access to procurement policies, solicitations, drawings and amendments. Vendors can browse the listings and register to receive automatic email notification of business opportunities.

Likewise, the GSA Schedules Program[31] serves as the catalyst for billions of dollars in Federal spending, helping meet procurement needs for eligible users, including all branches of Federal, state and local governments through applicable programs. Under the program, GSA establishes long-term, government-wide contracts with commercial firms to provide access to millions of commercial supplies (products) and services at volume discount pricing.

Since the 2006 National Biometrics *Challenge*, many competitive opportunities have been posted to FedBizOpps and technologies have been acquired through GSA Schedules, reflecting government efforts to fill the gaps and meet the challenges identified in 2006. The Intelligence Advanced Research Projects Activity (IARPA) Office of Smart Collections has initiated the Biometric Exploitation Science and Technology (BEST) Program to advance the state-of-the-science for biometrics

---

29 Martin Woodside, *Sterling Biographies: Thomas Edison: The Man Who Lit Up the World (2007)*: 17.
30 FedBizOpps, **www.FedBizOpps.gov**, accessed Aug. 9, 2011.
31 GSA, **www.gsa.gov**, accessed Aug. 9, 2011.

technologies significantly on behalf of the intelligence community. The IARPA BEST Program's overarching goals are to:

- Significantly advance the intelligence community's ability to achieve high-confidence match performance, even when the features are derived from non-ideal data; and

- Significantly relax the constraints currently required to acquire high-fidelity biometric signatures.

The IARPA BEST Program will consist of three phases over a five-year period. Phases 1 and 2 will be approximately 24 months each, and Phase 3 will be approximately 12 months. Multiple awards were provided for Phase 1 in late 2009; each Phase 1 award is envisioned to consist of a one-year base period with one option year. Near the conclusion of Phase 1, proposals for Phases 2 and 3 will be requested through the issuance of a separate solicitation.

The DOJ Office of Justice Programs releases several solicitations and grant programs each year. Most applicable to biometrics and identity management is the National Institute of Justice Sensors, Surveillance and Biometrics Technologies solicitation. Multiple awards are provided for novel sensor or surveillance technologies, applications or support functions for specific criminal justice applications. Along with providing several awards via various forensic solicitations, DOJ also facilitates the Community Oriented Policing Services (COPS) Program. COPS works principally by sharing information and making grants to police departments around the United States.

The DHS Science and Technology (S&T) Directorate, Human Factors and Behavioral Sciences Division (HF/BSD), periodically issues broad agency announcements and actively supports biometrics research and development. DHS S&T's high-level goals for specific biometric research areas correspond to the needs of its seven member agencies, and they are related to multimodal biometrics and mobile biometrics. In approaching these goals, DHS has funded research that focuses biometrics in new and existing DHS operations, developing technologies, sensors and components for integration in future multimodal mobile biometrics collection systems. DHS also has supported efforts to "improve screening by providing a science-based capability to identify known threats through accurate, timely and easy-to-use biometric identification and credentialing validation tools."[32]

Multiple groups within DOD have responsibility for biometrics research. The Biometrics Identity Management Agency (BIMA) serves as the proponent of biometrics within DOD. BIMA leads DOD activities in programming, integrating and synchronizing biometric technologies and capabilities and operating and maintaining DOD's authoritative biometric database to support the National Security Strategy. The Defense Advanced Research Projects Agency (DARPA) is the central research and development organization for DOD. DARPA's mission is to sponsor high-payoff research to uphold the American military's technological dominance, "bridging the gap between fundamental discoveries and their military use."[33]

Both organizations have made important contributions to biometrics research and development and are expected to continue to support future efforts.

Office of Personnel Management

On Oct. 14, 2009, at 12:16 p.m., IAFIS received an electronic fingerprint submission from the United States Office of Personnel Management for a pre-employment criminal background check. The fingerprints were processed on IAFIS. Within seconds, the individual was identified. The individual had previous arrests in San Antonia, Texas, for indecency with a child. At 12:16 p.m., a response containing this information was sent to the submitting and wanting agencies.

---

32 DHS, S&T HF/BSD, http://www.dhs.gov/xabout/structure/gc_1224537081868.shtm, accessed Aug. 9, 2011.
33 DARPA, http://www.darpa.mil/About.aspx, accessed Aug. 9, 2011.

Along with the previously mentioned broad agency announcements and solicitations, the large-scale biometrics and identity management systems including DOD ABIS, DHS IDENT, DOS CCD and FBI IAFIS/NGI, as well as state and local law enforcement agencies, have competed multi-million dollar development efforts, acquired sensor technologies and contracted subject-matter expertise to support their initiatives.

www.ingramcontent.com/pod-product-compliance
Lightning Source LLC
Chambersburg PA
CBHW081802170526
45167CB00008B/3296